# Liquid Crystals Through Experiments

# Liquid Crystals Through Experiments

**Mojca Čepič**

*University of Ljubljana, Slovenia*

Morgan & Claypool Publishers

ISBN 978-1-6270-5300-6 (ebook)
ISBN 9781643278377 (print)
ISBN 978-1-6270-5301-3 (mobi)

DOI 10.1088/978-1-6270-5300-6

Version: 20141201

IOP Concise Physics
ISSN 2053-2571 (online)
ISSN 2054-7307 (print)

A Morgan & Claypool publication as part of IOP Concise Physics
Published by Morgan & Claypool Publishers, 40 Oak Drive, San Rafael, CA, 94903, USA

IOP Publishing, Temple Circus, Temple Way, Bristol BS1 6HG, UK

*I would like to thank my very active research group, which currently comprises women only: Katarina, Jerneja, Saša and two Majas. The whole group was extremely effective in working as a team solving problems, and working with them was a real pleasure. Without them this textbook would be much thinner or, more likely, non-existent.*

*Although I tried to limit the work on this textbook to the time spent at the university, I was not always successful. Therefore many thanks to my partner Uroš and our children Maja and Blaž for their patience and understanding.*

# Contents

# Preface

Science subjects are often considered uninteresting, or boring, with no relevance to everyday life and experiences. On the contrary, scientists find their disciplines extremely interesting, working with constant curiosity and joy, and often describe their research as a passion. How is it that such differences in opinion exist between these two worlds? It seems that the first time students see relatively recent research results is at university when they enroll onto more advanced courses. Prior to this stage, the topics they have studied are mostly old. Theories of mechanics have been known for several hundred years, and the discoveries in thermodynamics, electromagnetism and other fields that are part of the teaching program are only a little more recent. The most recent advances are elementary ideas from modern physics, but even these are almost a hundred years old which, from a student's perspective, is still ancient.

Another problem relates to the scientific explanation of everyday experiences. The physics and chemistry of everyday life is complex. Usually several phenomena are intertwined and explanations of relatively common experiences such as friction and tyres, the physics and chemistry of cooking, lightning in a thunderstorm and many other examples, are not easy to explain at elementary levels. Therefore science subjects often fail to be relevant to everyday life for students.

It is difficult for teachers to follow new advances in research. Fields have become specialized and a lot of preliminary knowledge is necessary in order to understand new discoveries and to appreciate their importance. As a consequence, the teaching of science subjects can easily become even more detached from everyday experiences and new scientific results, not to mention their combination in applications to novel devices.

As a former high school teacher and a long term (more than twenty years) lecturer of physics and the didactics of physics for future teachers of physics, I am interested in the problem of introducing a new, scientifically interesting, actively researched topic, and this has been my motivation for several years. As I have actively worked on the theoretical physics of liquid crystals for almost thirty years already, the choice of topic was straightforward. In the 1990s, the first simple experiments presenting basic properties to students and teachers were designed and were well accepted by motivated high school teachers. Although teachers have found the topic interesting, problems have occurred with teachers not having the necessary knowledge to competently teach it, not having the resources to acquire the necessary equipment and the placement of the topic in a relatively rigid curriculum. Nevertheless, the development of the experiments was interesting in itself, and as PhD students joined the project, the idea of developing a set of experiments that illustrate the complexity and peculiarity of the liquid crystalline world has become more feasible. The four PhD students have specialized in the different subtopics: Dr Katarina Susman studied mechanical models of phase transitions, Dr Saša Ziherl developed the whole set of experiments illustrating optics of anisotropic materials by microwaves and wood, Dr Jerneja Pavlin developed and tested a teaching module on liquid crystals and Dr Maja Pečar developed the set of experiments introducing a complex method, very often used in liquid crystals, the conoscopy. The whole group was extremely effective in working as a team to solve the problems, and working with them was a real pleasure.

Furthermore, the introduction of these topics at university level is not as limited by a strict curriculum.

The textbook is organized as follows. At the beginning, in chapter 1, a very short description of liquid crystals is given. It explains elementary ideas about the aims of the experiments, and motivates the reader to read some textbooks about liquid crystals from various other authors. The experiments are presented in three chapters. Chapter 2 presents a wide set of experiments that use liquid crystals as a material. Some experiments are very simple, others are more complicated, several of them could be developed into laboratory exercises in advanced labs and into projects. The following chapter, chapter 3, is shorter and focuses on the liquid crystal display, its structure and its function. For the sake of completeness, colour printing is also discussed. The last experimental chapter, chapter 4, presents several experiments that offer additional insights into phenomena which require comprehension in order to understand phenomena in liquid crystals. The attraction of these experiments is that they use easily accessible and easily manipulated materials, like tapes and cloth, and they provide easily observable evidence such as microwaves in an anisotropic wood. As the whole textbook aims to stimulate the introduction of the topic of liquid crystals to high school and university levels, the idea of such an introduction was tested too. A short report on the structure of the teaching module and its evaluation is given in chapter 5.

Finally, for readers happy to discover a new topic to teach or for teachers willing to go further in developing ideas about experiments using liquid crystals, a short overview of published papers related to liquid crystals in education is given in chapter 6. Several papers have been published in journals discussing university level education, and many of them can be developed further into interesting but demanding experiments.

Experimental chapters are organized in the following way. The chapters are divided to sections covering phenomena with the same origin. Specific experiments are presented in subsections if there are a number of them. The phenomenon illustrated by an experiment is presented and discussed at the beginning. The aims of a specific experiment are quoted and then the list of necessary equipment and materials is given. Instructions for the experiment follows and are illustrated where possible by photos of setups, the geometry for the analysis of results and examples of expected results. Most of the results are those obtained by our group when the experiments were performed during their development. When problems in the execution of an experiment can be expected, comments called *Tricks of the trade* are added. In cases when our experience has shown that students might develop misconceptions *Suggestions for the lecturer* are given. Finally, almost all experiments end with **Additional tasks**. Sometimes tasks are simply derivations of equations, calculations of expected results for measurements or the extraction of physical quantities from the measured data. Very often tasks are additional experiments using the same equipment as presented in the subsection and could be further developed into projects.

Ljubljana, 8 September 2014 — Mojca Čepič

# Acknowledgements

First I would like to acknowledge the work of people who have changed the English of a non-native, inexperienced writer, into something much better and enjoyable to read.

I would also like to acknowledge two projects financed by the Slovenian Agency for Science and Development, J5–0365 Teaching and learning of complex interdisciplinary topics in physics, and J5–4002 Introduction of current research interdisciplinary topic into education – liquid crystals.

Finally, the financial support from the Slovenian Agency for Science and Development for three PhD students Dr Katarina Susman, Dr Saša Ziherl and Dr Maja Pečar is acknowledged.

# Author biography

## Mojca Čepič

Professor Dr Mojca Čepič started her career as a high school teacher of physics. After a few years she left the school and became a PhD student in a program financed by the Slovenian Agency for Science and Development called Young researchers and graduated in theoretical studies of soft matter physics, more precisely, by development of a phenomenological theoretical model describing phases in antiferroelectric liquid crystals at the Faculty of Mathematics and Physics, University of Ljubljana, Slovenia. After finishing her PhD she started to work as an assistant for physics at the Faculty of Education, University of Ljubljana, where she has returned to teaching physics.

Today she is the lecturer of physics, physics education and science education within university programs for future teachers of physics, teachers of science and for primary school teachers.

She is still actively involved in theoretical soft matter physics, mainly in liquid crystals and has published over 70 articles in this field; several papers in co-authorship were published in *Physical Review Letters* and a review paper on antiferroelectric liquid crystals was published in *Reviews of Modern Physics*.

Soon after she became involved in the didactics of physics, she realized that the transfer of new discoveries in physics into schools and to undergraduate programs is almost non-existent. Such an introduction is difficult as students' knowledge is usually too basic to allow them to easily understand newly discovered phenomena. Therefore she has started to construct simple, more or less hands-on experiments that reflect new research results and allow students to have personal experience and obtain new knowledge emphasizing concepts important for the physics of liquid crystals. She has become increasingly active in the field of physics education and in research in physics education. During the last few years she has had more than 20 contributions published in journals that consider physics education, she has run workshops on active learning about liquid crystals in several European locations (as well as Slovenia also Serbia, Finland, Turkey, Czech Republic and Slovakia) in Mexico and America. Several experiments directly related to liquid crystals and several experiments illustrating concepts important for understanding liquid crystals that were developed with the purpose of introducing these beautiful new materials into education, are presented in the textbook Liquid Crystals Through Experiments.

Liquid Crystals Through Experiments

Mojca Čepič

# Chapter 1

## Liquid crystals

The first of the experimental chapters, chapter 2, presents experiments that demonstrate a small part of a rich phenomena characteristic of liquid crystals using liquid crystals themselves. These experiments represent the main part of the textbook. All of the experiments can be performed by simple means and are accessible to any modestly equipped school. Most of the experiments are divided into two parts. The first part is an essential observation and is meant for demonstration purposes during the lecture or for providing elementary experiences in the introduction of students to laboratory work. Several activities are then elaborated on by more detailed measurements. Measurements remain at the school laboratory level, therefore equipment is simple and accuracy of measurement is not the main goal. Nevertheless, methods used in the presented experiments are actually equal to methods used in research. The only difference is that experimental equipment is as simple as possible and it provides insight into the operation of various instruments used in research laboratories wherever possible. The reason for insisting on simple hands-on experimental equipment is twofold. Firstly, it is hoped that the textbook will be used in undergraduate courses where concepts are introduced and sophisticated experimental equipment typical in research laboratories is not available. Secondly, professional instruments often act as black boxes, especially when students are at the beginning of their research career. Preliminary experiences with much simpler, though less accurate, experimental equipment with the same function as professional versions provide insight into the physics concepts that experimental methods are based on. An example presented in figure 1.1 of a simple observation of a liquid crystal between crossed polarizers, and a polarizing microscope usually used in a research laboratory, shows the difference between the two. The chapter is organized as follows. It starts with a short description of liquid crystals as materials that have additional phase. Reasons for the special properties are presented through the concept of an order parameter and how the order parameter is related to the anisotropy in dielectric and optical properties.

This chapter is a general description of materials, their chemical properties such as general structures and some elementary formulas, and their physical properties,

doi:10.1088/978-1-6270-5300-6ch1  

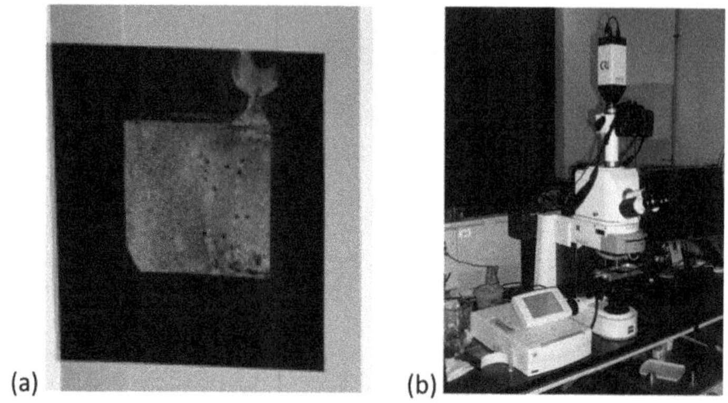

**Figure 1.1.** (*a*) A liquid crystal in the cell between crossed polarizers (Photo: Maja Pečar) and (*b*) the professional polarizing microscope, which allow essentially very similar observation (Photo: Damian Pociecha).

such as phases, their structures at the microscopic level and their optical and electrical properties. The chapter is aimed at a general reader who is an expert in other fields in physics and wants to get a general idea about liquid crystals, either to include some experiments in lectures in optics or to include some illustrations during lectures in soft matter. Liquid crystals are easily included in general undergraduate lectures to provide information about ongoing research. The physics knowledge needed to understand the basics is not advanced; an elementary knowledge of phase transitions, electric polarization and electromagnetic waves is enough to enable a student to understand the phenomena related to liquid crystals and their special properties.

Liquid crystals are organic materials which have at least one additional state of matter between the liquid phase and the solid phase [1]. Between the two phases, the transparent liquid phase and the powder-like crystal solid phase, an additional phase, sometimes with several subphases, appears. If the liquid crystal is observed in a test tube, the transparent liquid becomes opaque upon cooling. The opaqueness remains almost unchanged for several degrees and the material freezes into powder-like crystal upon further cooling. The opaque liquid reveals different viscosity to the transparent isotropic phase, additional evidence for a different phase. That different phases often exist in these materials is best seen if a drop of a material is put on a clean slide, covered with another slide and observed between crossed polarizers. In the temperature range where it is opaque, the material is transparent, often coloured, and reveals several features between crossed polarizers such as dark threads, variable colouring and similar (see figure 1.2). The colours and transparency of the material between crossed polarizers is an irrefutable indication of anisotropy that is typical for crystals. This state of matter is therefore called the liquid crystalline phase because the material in this phase flows like liquids and is anisotropic like crystals. A whole zoo of various liquid crystalline phases with a large variety of properties is associated with a general name—the liquid crystalline phase. Sometimes many of them are found in a single material. The name liquid crystals is reserved for materials which possess at least one of the liquid crystalline phases.

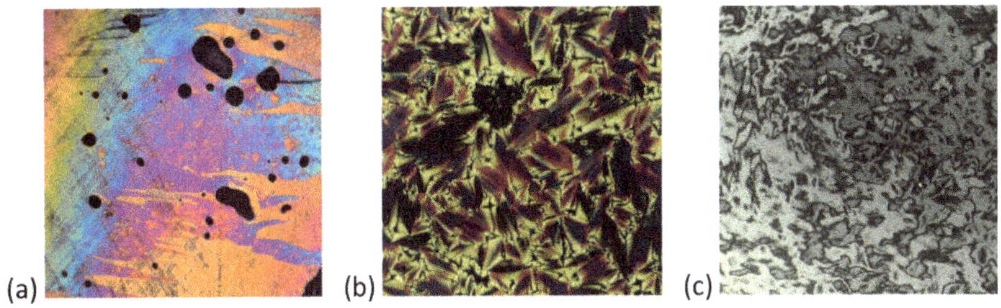

**Figure 1.2.** A thin layer of a liquid crystal in a cell between crossed polarizers under the microscope (*a*) in the nematic phase (Photo: Maja Pečar), (*b*) in the smectic phase (Photo: Damian Pociecha), (*c*) in the polar smectic C phase (Photo: Damian Pociecha).

## 1.1 Chemical properties

Liquid crystals are organic materials. Several benzene rings are connected linearly giving the molecule an elongated form. This part of a molecule is called the *molecular core*. Alkyl chains are attached to both sides of the molecule. They are called the *flexible tails*. Molecules described here typically have one dimension much larger then the other two, usually around 3 nm relating to the length and 0.5 nm relating to the width of the molecule. Therefore the molecules are elongated and are colloquially called rod-like molecules. Both the non-flexible molecular cores and the flexible tails give the molecules a different mobility and consequently the viscosity of such materials is different (higher) than of materials consisting of more symmetric molecules. As some types of molecules have attached chiral groups and oxygens, which contribute to molecular dipolar moments, several different intermolecular interactions exist and result in various liquid crystalline phases that will be described in more detail later. A few chemical formulas of typical representatives of liquid crystals are found in table 1.1 accompanied by their phase sequences.

Another sort of molecule which forms liquid crystals are disc-like molecules. They also consist of several benzene rings connected to a rigid core, however they are not attached linearly to one another but form a disc-like core structure laterally. Flexible tails are attached to the benzenes giving a soft flexible halo around the disc. The molecular structure has given them their name as they are much shorter in one dimension and wider in the other two. Such molecules form different structures to the rod-like molecules and also have different properties. Two examples are given in table 1.2. We shall not discuss the properties of liquid crystals formed from disc-like molecules although they are academically motivating due to their interesting physics. They are also not easily accessible for non-experts. Nevertheless, we will mention differences in properties between rod-like liquid crystals and disc-like liquid crystals where appropriate. The application of experiences and knowledge that students gain through these experiments will allow students to also understand the properties of discotic liquid crystals.

## 1.2 Physical properties

The special shapes of molecules result in different attractive interactions among the different parts of the molecules. The core of a molecule interacts with the core of

**Table 1.1.** Chemical structure for some representative examples of rod-like molecules forming liquid crystals with corresponding phase sequences upon cooling and transition temperatures.

| Material short name | Chemical formula | Phase sequence |
|---|---|---|
| MBBA | 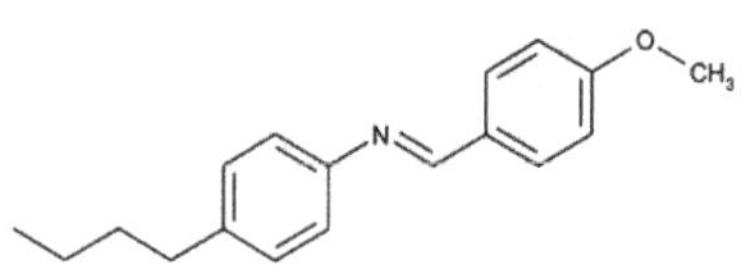 | I ↔ N |
| 5CB | 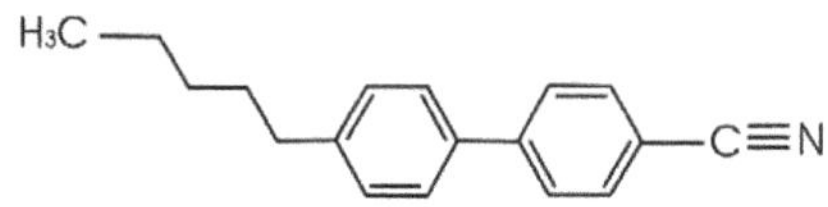 | I ↔ N ↔ SmA |
| cholesteryl benzoate | 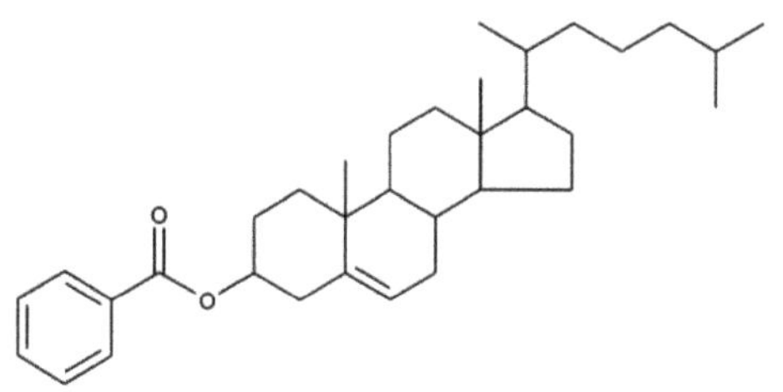 | I ↔ N* |
| DOBAMC | 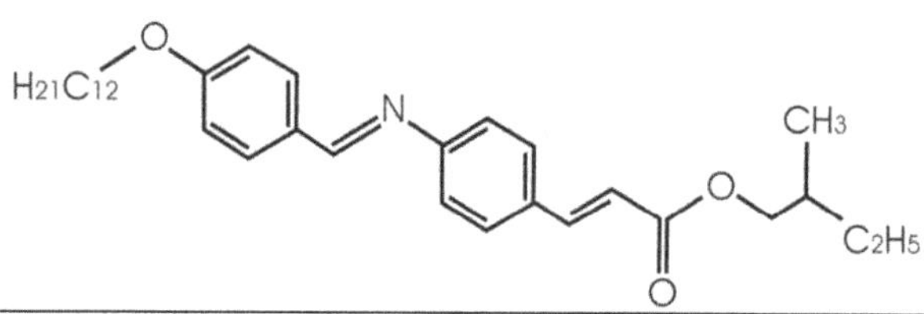 | SmA ↔ Sm $C_{\alpha}^{*}$ |

another molecule more strongly than with its own tail. Interactions also vary with temperature changes and therefore different types of orders are stabilized at different temperatures (see figure 1.2).

The crystalline state becomes stable when van der Waals attractive forces prevail over the entropically desired motion of molecules. The opposite is true for the isotropic liquid phase. In liquid crystals another type of ordering occurs, balancing the 'competing actions' of entropically favoured movement and van der Waals attraction, leading to favourable fixed positions in places where interaction bonds between the molecules are the strongest. If elongated molecules orient themselves parallel to each other, the distances between parts of molecules are shorter in comparison to other mutual orientations of molecules in the group, and the parallel orientation is therefore favoured by van der Waals attraction. However, if molecules move one parallel to another along their long axes, distances between parts of molecules remain approximately the same, but movement, that is diffusion, satisfies tendencies

**Table 1.2.** Chemical structure for some representative examples of disc-like molecules forming liquid crystals.

| Material short name | Chemical formula |
| --- | --- |
| hexaazotriphenylene | 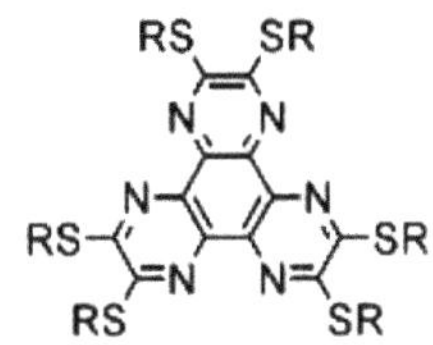 |
| phthalocyanine | 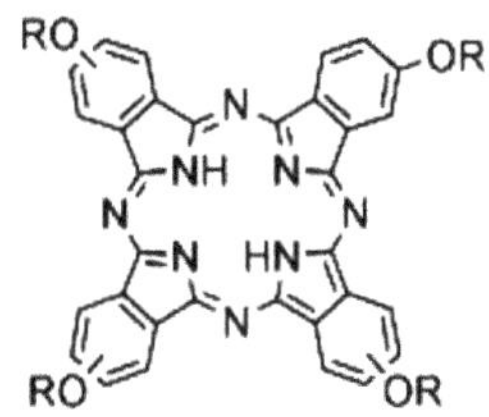 |

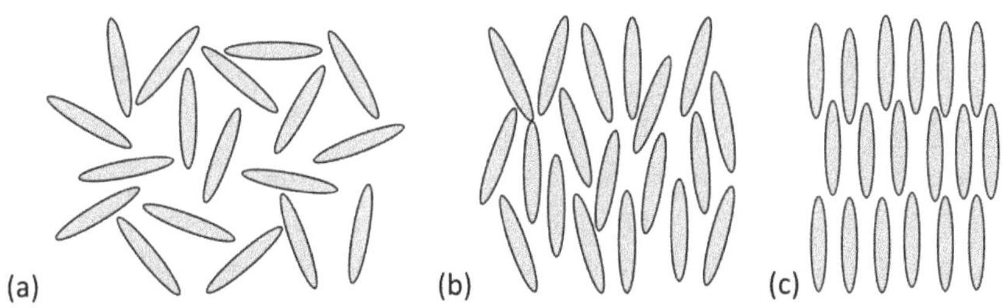


**Figure 1.3.** (*a*) Schematic diagram of the isotropic phase. (*b*) Schematic diagram of the nematic phase. (*c*) Schematic diagram of the crystalline phase.

for entropic disorder. Therefore the structure where molecules are orientationally ordered but still move compensates entropy of diffusion with higher order in orientation and the new phase becomes stable—the nematic liquid crystalline phase.

The nematic phase is schematically presented in figure 1.3(*b*). The figure clearly shows why we were discussing distances between parts of different molecules. Typical molecules forming a liquid crystal have lengths of approximately 3 nm and widths of about 0.5 nm, so the length to width ratio is around 6:1. Intermolecular distances in the nematic phase are much shorter than the molecular lengths. The structure of the nematic phase is closer to that of a crystal, as molecules are orientationally ordered, that is the order is higher than in the isotropic liquid phase and the material in the nematic phase is anisotropically similar to crystals. However, molecules are still able to move, which gives the material properties characteristic of liquids. The name 'liquid crystal' comprises both the fluidity and the anisotropy. In the nematic phase molecules are not positionally ordered, their long molecular axes are predominantly oriented in one direction only. However, in the nematic phase molecules are positionally disordered and their positions have liquid-like distribution. A distribution of orientations of molecular long axes is centred around a specific direction and are *orientationally ordered.* Molecules are, on average, oriented in the same direction, and they are orientationally ordered over a part of a volume that has dimensions of about several hundreds of μm. Liquid crystals that have a wide temperature range for their nematic phase are mostly used in liquid crystal displays.

If molecules are chiral, elongated molecules are not parallel like in figure 1.3(*b*) but prefer a local twist. The sense of the twist depends on molecular chirality. Such phases are called the *chiral nematic* phases [2]. They are used in liquid crystalline thermometers where colour changes indicate the temperature changes. Another series of very complex phases may form with chiral elongated molecules. They have a common name, *blue phases,* due to the blue colour of the reflected light in the first observations of these phases. In blue phases the chiral twist develops in various directions in space and the final structure is rather complicated [3]. These phases were for a long time considered purely academically interesting, but recently there have been several suggestions for possible applications in photonics [4].

In some materials the structure changes again upon further cooling. The difference in interactions among cores and among tails leads to the separation of cores and tails and results in another type of liquid crystalline phase—the *smectic* liquid crystalline phase. In the most simple smectic phase, called the smectic A phase, the elongated molecules are ordered orientationally and, in addition, they are organized in smectic layers, see figure 1.4(*a*). On average, the long molecular axes are parallel to the layer normal. A one dimensional pattern of cores and tails is established. The appearance of the smectic phase in a test tube is the same as the appearance of a nematic liquid crystal, but the viscosity is higher. The high viscosity is reasonable as smectic layers have to be broken during the flow. Smectic phases have several subphases that differ in the intralayer organization of molecules. Some subphases have tilted long molecular axes, see figure 1.4(*b*), some have different types of intralayer positional correlations, and so on. If molecules that have a tendency to form layers are chiral, the whole zoo of new phases appears. Smectic layers are broken at various places and at those places a chiral nematic structure exists [5, 6]. A network of defects is formed in a regular or in an irregular way. However, if chiral molecules form layers and are also tilted away from a layer normal, the layers are polar.

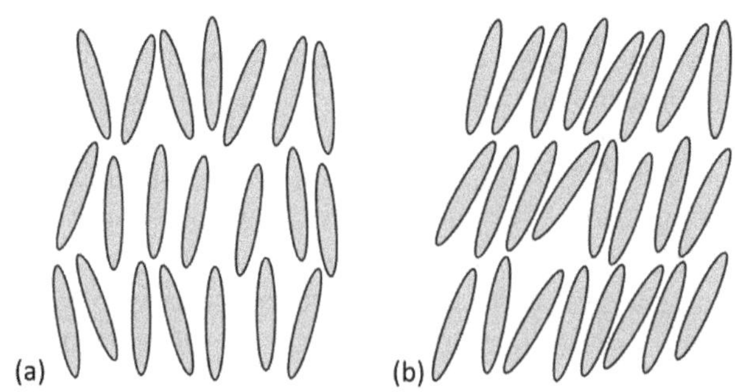


**Figure 1.4.** (*a*) Schematic diagram of orthogonal smectic A phase. (*b*) Schematic diagram of tilted smectic C phase.

Such materials are very sensitive to an external electric field, due to polarity. As polar electrostatic interactions add a new dimension to intermolecular interactions, the structures formed in such materials have several additional modulations of longer and shorter range and several structurally different subphases exist. There have been several applications suggested for such materials but the devices have not yet been developed far enough to be on the market. Collaborations of researchers in chemistry and physics have resulted in several new materials with complex structures that are revealed in the complexity of the phase structures and corresponding phase diagrams. Studies of liquid crystals from various aspects are very interesting. More detail about phases, research results and similar, can be found in many good textbooks, for example [7, 8].

### 1.2.1 Order in liquid crystals

In this textbook, we mainly consider liquid crystals formed of elongated molecules that order under certain conditions with long axes in the same direction but remain positionally uncorrelated and able to move. As a consequence, anisotropic properties appear. Anisotropy exists in electric susceptibility and permittivity, magnetic susceptibility and permeability, viscosity, and other properties. All of them have a tensorial character and they are all determined by the state of order. In addition, they also depend on molecular properties and are therefore not appropriate as a measure of order. For a proper analysis of the liquid crystalline order, a measure independent of molecular properties is needed. There are several different phases in liquid crystals and they differ in structure and symmetry. For example, nematic liquid crystals have only orientational order, in smectic A liquid crystals molecules order into layers in addition to nematic ordering, in smectic C liquid crystals order in a different way (the nematic director becomes tilted with respect to the layer normal), and if molecules are chiral, that is they have no mirror symmetry, layers become polar. Ordering of dipoles presents a new type of order, the polar order. For any of these specifics, a new order parameter is introduced that has to reflect the structure and the symmetry of the order and should be related to macroscopic properties of materials. For example, the nematic order parameter has to reflect the up–down symmetry, because in the nematic phase there are as many molecules oriented in one direction as there are in the opposite direction. That is to say, there is no polar order. The smectic order has to measure the quality of the smectic layer formation and is a simple scalar added to nematic order. Both are needed to describe the state of order in a smectic A phase. In the smectic C phase the third order parameter is needed, which is related to the magnitude of the average tilt of long molecular axes. In polar liquid crystals the ordering of molecular dipoles should be described as well, therefore one needs one additional new order parameter, and so on. The more complex the phase, the greater the number of different order parameters that are needed to describe the order.

As in all the experiments that use liquid crystals, only the nematic phase will be considered; we shall focus on a nematic order parameter in more details. Let us consider a part of a volume that contains a large number of molecules. They are on average oriented in the same direction, but this direction is arbitrary. The average orientation of long molecular axes of a cluster is called a *director* and it is represented by a unit vector in the average direction. Note that two directors oriented in exactly opposite directions present the same orientation due to the up–down (quadrupolar) symmetry of the nematic phase. When the structure of the nematic liquid crystal in a cell or in a defect is presented, usually a short line indicating molecular orientation is used to avoid this problem. The director also does not give any information about the order itself, whether the order is high or low. A semi-quantitative comparison can be made when comparing two structures. One can easily recognize that in figure 1.5(*a*) the order is lower than in figure 1.5(*b*), although directors marked as a bold line at the edge are the same in both cases. Therefore one should define the procedure to express the order in numbers. What are the properties of a properly designed order parameter?

- It should be independent of the direction in which elongated molecules orient in the space on average.
- It should reflect the symmetry of the phase. For nematic liquid crystals, the order parameter should not change under rotation of each molecule for 180°.
- It should be zero in disordered systems.
- It should be one in an ideally ordered system, where all molecules are oriented in exactly the same direction.

Molecules are oriented in the same direction on average, along the director, but a focus on one molecule shows that a specific molecule is most probably not oriented along the director but forms an angle with it. In addition, it librates and changes its orientation in time. The probability of finding the specific angle that a long molecular axis forms with a director depends on the angular distribution of the molecules that are centred around the director direction. The first idea is that the order parameter has to be related to this distribution. If the probability of finding a long molecular axis of a specific molecule forming large angles with the director is relatively high, the order is probably low. However, if all molecules are oriented strictly along the director, the highest possible order is achieved. Let us associate a *molecular director* with an instant orientation of the *i*th molecular long axis. Consider the angle $\theta_i$ that is an angle between a director $n$ and the molecular director $n_i$. As a measure of order, one would think about averaging projections of molecular directors

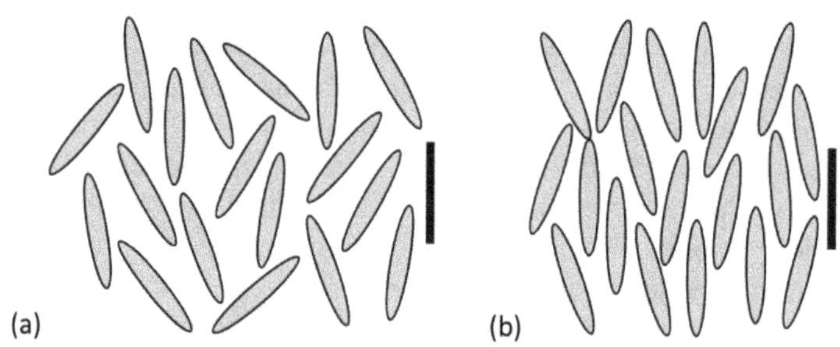


**Figure 1.5.** (*a*) Schematic diagram of a nematic phase with a low order parameter. (*b*) Schematic diagram of a nematic phase with a high order parameter. The director is marked on the right hand side on both diagrams and is the same.

$n_i$ to the director $n$ over all molecules. In this way, the problem of arbitrary average direction of long molecular axes that should not reflect in the order parameter is solved. However, averaging the angle $\theta_i$ over all molecules would result in zero for any molecular distribution because of the quadrupolar symmetry of the phase. So the average projection itself is not an appropriate measure. The second guess would be a square of the component, which has a proper quadrupolar symmetry, but a thermodynamical average of this quantity is not zero in a disordered phase.

$$\langle \cos^2 \theta_i \rangle = \frac{1}{3} \tag{1.1}$$

A renormalization solves the problem. The $\frac{1}{3}$ is subtracted from the average giving the required zero for a disordered phase. Now, the order parameter is equal to $\frac{2}{3}$ in an ideally ordered phase. The multiplication by $\frac{3}{2}$ solves this problem as well. The nematic order parameter $S$ is defined as

$$S = \frac{3}{2}\left(\langle \cos^2 \theta_i \rangle - \frac{1}{3}\right). \tag{1.2}$$

The order parameter changes with temperature. A typical dependence of order parameter in a nematic liquid crystal is given in figure 1.6($a$). From zero in an isotropic phase, it jumps to a certain value of around 0.3 at the clearing temperature, and gradually grows upon lowering the temperature.

There is no direct method to measure an order parameter. It is usually deduced from birefringence, see figure 1.6($b$), from a diamagnetic anisotropy or other properties that are proportional to the order parameter in the first approximation. More details about the orientational distribution of molecules are extracted from the x-ray measurements, but a detailed description of the method goes beyond the purposes of

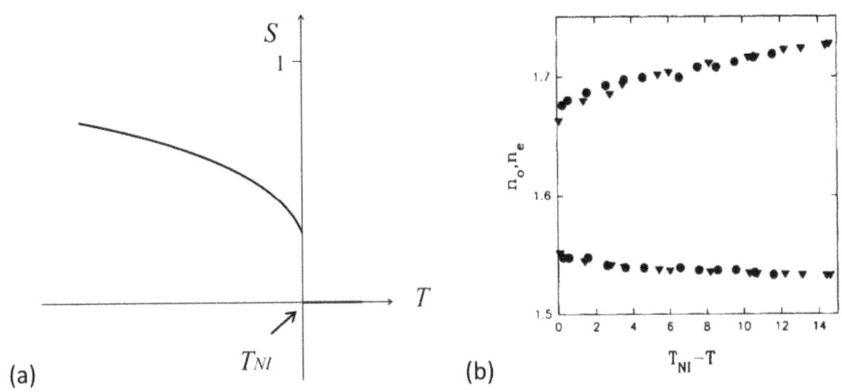


**Figure 1.6.** ($a$) The dependence of an order parameter on the temperature (after [9]). ($b$) The dependence of the two refractive indices (for polarization parallel to the directors $n_e$ and perpendicular to it $n_o$) on temperature [10]. Note that the birefringence is a difference between the ordinary and extraordinary refractive index and its dependence on temperature is very similar to the dependence of the order parameter.

this textbook. The strict definition of the order parameter is used in computer simulations where detailed dynamics of a large set of molecules is simulated. As one knows the positions and orientations of all modelled molecules in the simulated sample, the procedure described above can be applied according to definition.

The phase transition from an isotropic to the nematic state is always discontinuous. The order parameter jumps from zero to the value around 0.3 upon cooling. Other transitions, from nematic to smectic, for example, can be continuous or discontinuous. Therefore liquid crystals present a good example of materials where the order of a transition can be discussed as well, and can be used as a good illustration in courses on thermodynamics. Unfortunately, experiments indicating the order of transition are not simple and are beyond the scope of this book. We intend to show a few phenomena associated with discontinuous transitions only.

### 1.2.2 Dielectric properties

Molecules that form liquid crystals are highly anisotropic, and either elongated or disc-like in their generalized shape. In the continuation we shall discuss properties of systems formed of elongated molecules as they are the main concern in the experiments presented later. If such elongated molecules are found in an electric field their polarization depends on their orientation with respect to the external electric field. An induced polarization is usually the largest if the electric field is parallel to the long molecular axes and it is the smallest if the electric field is perpendicular to the long molecular axes. However, for some materials the opposite is true. If molecules are ordered, the polarizability of the ordered media is reflected in an electric susceptibility $\chi$ that is a tensor.

$$\underline{\chi} = \begin{bmatrix} \chi_{\parallel} & 0 & 0 \\ 0 & \chi_{\perp} & 0 \\ 0 & 0 & \chi_{\perp} \end{bmatrix} \tag{1.3}$$

Here $\chi_{\parallel}$ is the polarizability for an electric field parallel to the director, and $\chi_{\perp}$ is the polarizability perpendicular to it.

What actually happens in an electric field? An anisotropic molecule becomes polarized. If the molecule is not oriented in one of the two specific directions with respect to the electric field, perpendicular or parallel to it, the torque of the electric field is applied to the induced molecular dipole and consequently to the molecule (figure 1.7).

$$\begin{aligned} \vec{P} &= \epsilon_0 \underline{\chi} \vec{E} \\ \vec{M} &= -\vec{P} \cdot \vec{E} = -\left(\epsilon_0 \underline{\chi} \vec{E}\right) \cdot \vec{E} \end{aligned} \tag{1.4}$$

The molecules tend to align parallel or perpendicular to the electric field, depending on the sign of a dielectric anisotropy, but they are prevented by an entropic 'desire' for rotational and translational diffusion, intermolecular interactions or interactions with surfaces. Molecular aligning in an external electric field is small in an isotropic, disordered phase. In a disordered phase the orientational diffusion is much stronger

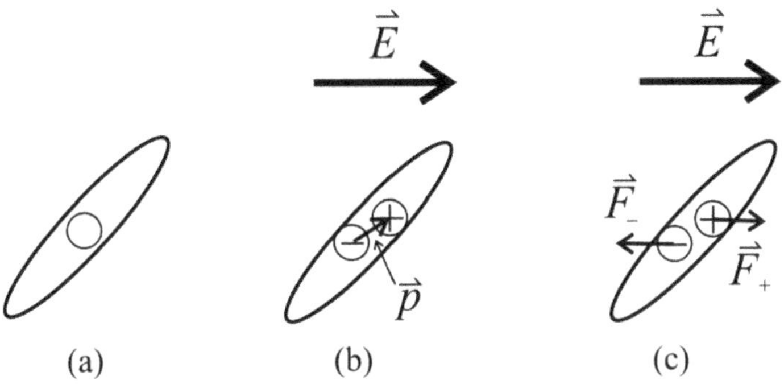


**Figure 1.7.** (*a*) Non-polar molecule. (*b*) External electric field induces polarization. Note that the induced dipole is neither parallel to the external electric field nor to the long axis of the molecule. Susceptibility is a tensor. (*c*) Induced polarization interacts with an external electric field and the torque tries to align the long molecular axis parallel to the external electric field. Note the mirror symmetry of the phenomenon as well.

than at lower temperatures where molecules are ordered to one direction predominantly. In the ordered nematic phase, the 'softness' of the material is manifested. The orientation of molecules is arbitrary in general but weak interactions with surfaces of the cell is already enough to determine the general orientation direction of molecules. However, when the field is applied, the molecules that are used in a liquid crystalline screen, for example, tend to orient with long molecular axes along the electric field. The competition between surface interactions and torques of the electric field applied to induced dipoles result in a non-homogeneous director across the cell. Close to the surfaces, interaction with surfaces prevail and molecules are predominantly oriented as determined by the surface treatment. At larger distances from surfaces, molecules are oriented more toward the electric field. The reorientation in the middle of the cell starts at a certain critical electric field and has all the characteristics of the structural transition. By structural transition, it is meant that a change in the structure occurs due to external influences such as an electric or a magnetic field and not due to the temperature, which is associated with phase transitions. This structural transition is call the Freederickzs transition and it is the basis of the liquid crystalline display [11].

### 1.2.3 Optical properties

Anisotropy in optical properties is the essence of all liquid crystal displays (LCDs), therefore it is worth discussing. Optical properties depend on dielectric properties in a frequency range of electromagnetic waves corresponding to visible light, that is from $4.3 \times 10^{14}\ \mathrm{s}^{-1}$ to $7.5 \times 10^{14}\ \mathrm{s}^{-1}$. In this range, typical electrical permittivities are from 2 to 3.5, resulting in refractive indices from 1.5 to 1.8. These values result in typical birefringences, the differences between the refractive indices for linearly polarized light along the director, and perpendicular to it, of around 0.3. Such birefringence is huge and allows us to exploit the effects of birefringence on polarization states of light for the very thin layers necessary for good order of liquid crystals, and for the relatively high electric fields obtained by voltages of only a few volts. The typical thickness of a liquid crystal in a screen is a few μm, that is, far below the director correlation length and it guarantees good ordering. However, this thickness corresponds to 10 to 15 wavelengths, depending on the colour of the transmitted light. The thickness of the liquid crystal in a display is large in comparison to the wavelength of light, and the molecular arrangement in the cell strongly influences the polarization state of the light. If the long molecular axes remain perpendicular to the direction of light propagation, but the director continuously rotates from one surface to another, the changes are slow enough that the polarization of light being parallel to long molecular axes at the glass surface on the side of incidence also rotates in unison with the rotation of the molecular axes. However, these effects are not so strong if molecules reorient toward the electric field normal to liquid crystal display glasses. For light propagating through the display, the difference in indices for polarizations not parallel to the long molecular axes is smaller. The rotation is not perfect but light becomes elliptically polarized and the absorption is larger. In this way, grey levels are achieved.

So, the basis of the LCD is the change of molecular arrangement in the screen upon application of electric field. By addressing different positions in the LCD with different electric fields, that is, voltages, the local arrangement changes, the intensity of transmitted light varies from one place on the screen to another, and the image on the screen appears. As the most important device based on liquid crystals, the structure and function of the liquid crystalline display will be presented in detail in chapter 3.

## Bibliography

[1] Collings P J 2002 *Liquid Crystals: Nature's Delicate Phase of Matter* (Princeton, NJ: Princeton University Press)

[2] Oswald P and Pieranski P 2005 *Nematic and Cholesterid Liquid Crystals: Concepts and Physical Properties Illustrated by Experiments* Liquid Crystals Book Series (London: Taylor & Francis)

[3] Wright D C and Mermin N D 1989 Crystalline liquids: the blue phases *Rev. Mod. Phys.* **61** 385–432

[4] Kikuchi H, Yokota M, Hisakado Y, Yang H and Kajiyama T 2002 Polymer-stabilized liquid crystal blue phases *Nature Materials* **1** 64–68

[5] Renn S R and Lubensky T C 1988 *Phy. Rev. A* **38**(4) 2132–47

[6] Matsumoto E A, Alexander G P and Kamien R D 2009 Helical Nanofilaments and the High Chirality Limit of Smectics A *Phys. Rev. Lett.* **103**(25) 257804–1–4

[7] Oswald P and Pieranski P 2005 *Smectic and Columnar Liquid Crystals: Concepts and Physical Properties Illustrated by Experiments.* Liquid Crystals Book Series (London: Taylor & Francis)

[8] Goodby J W, Collings P J, Kato T, Tschierske C, Gleeson H and Raynes P (ed) 2014 *Handbook of Liquid Crystals; 8 Volumes* 2nd edn (New York: Wiley-VCH)

[9] de Gennes P G and Prost J 1995 *The Physics of Liquid Crystals.* International Series of Monographs on Physics (Oxford: Clarendon Press)

[10] Shenoy K D 1994 Measurement of liquid crystal refractive indices *Am. J. Phys.* **62** 858–9

[11] Moses T 1998 The Frèedericksz transition in liquid crystals: An undergraduate experiment for the advanced laboratory *Am. J. Phy.* **66**(1) 49–56

# Chapter 2

# Experiments with liquid crystals

This chapter suggests and discusses experiments that illustrate and study phenomena in liquid crystals. The chapter starts with the synthesis of two liquid crystals; however, in most of the chapter only the product of the first suggested synthesis, the MBBA, is used. The lecturer should use other liquid crystals that have a nematic phase which is stable around room temperature. Actually, it is advised to use other liquid crystals from providers due to the unpleasant smell of MBBA.

The experiments cover all aspects relevant for introductory physics and optics courses. The simplicity and the hands-on approach is emphasized. The equipment needed is simple, but experiments provide valuable experiences for later graduate studies. Several experiments could be extended or used in their suggested form as laboratory exercises or as projects for students.

## 2.1 Synthesis

Two examples of liquid crystals are relatively easy to synthesize in a modest school laboratory. One is a liquid crystal that exhibits the nematic phase N-(4-metoxibenzylidylen)-4-butylanylin, also called MBBA, the other one is the cholesteric or chiral nematic liquid crystal named cholesteril benzoat. Here, both syntheses are quickly described, but more detail can be found in the original papers [1, 2].

### 2.1.1 MBBA

Liquid crystal MBBA, N-(4-metoxibenzylidylen)-4-butylanylin, was the first liquid crystal to exhibit a nematic phase at room temperature. It belongs to a group of nematic liquid crystals, which are the simplest liquid crystals to synthesize and is most useful for teaching purposes, from various aspects. Details about the synthesis are given in figure 2.1 and were published in [1]; here the procedure is only summarized. The synthesis is easy and is successful even for students without any experience in chemistry. The synthesis takes around 90 min. There is some spare time due to the rinsing of the liquid crystal, so it is a good idea to combine the

doi:10.1088/978-1-6270-5300-6ch2

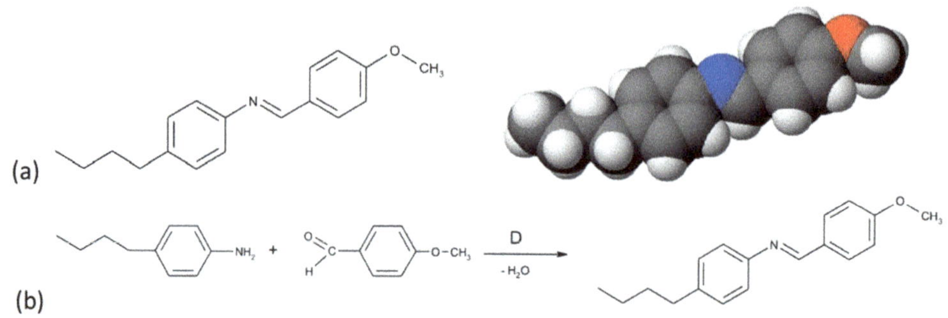

**Figure 2.1.** (*a*) The chemical formula and a model of an MBBA molecule. (*b*) The synthesis of MBBA.

synthesis of a liquid crystal with some other activities. The only problem for pre-university institutions could be that a fume hood is necessary.

The mixture of 4-butylaniline p-anisaldehyde in the ratio 1.48:1.37 is carefully heated for 20 min. The reaction is checked from time to time by putting a small drop of the product on glass to cool it down to room temperature. It should become cloudy after a few seconds. This is the sign of a successful synthesis. The mixture is then cleared by toluene and an aqueous solution of acetic acid. The rinsing is repeated several times. Transition temperatures are measured in a water bath. For educational experiments it is enough if the melting temperature is close to 20 °C and the clearing temperature close to 40 °C. The refined product is prepared to be used in experiments.

### 2.1.2 Cholesteryl benzoat

Cholesteryl benzoat is a liquid crystal that was discovered in 1888 by Friedrich Reinitzer, who was the first to discover that a material has something similar to two different phase transitions. At the first transition at 145.5 °C the material becomes an opaque liquid. When heated further, it becomes clear at 178.5 °C. The intermediate phase was later called a liquid crystal [3].

The synthesis follows the steps given in figure 2.2. Cholesterol and benzoyl chloride in the ratio 9.7:4.2 are dissolved in pyridine and refluxed for 3–5 min. The solution is allowed to cool slowly and the product is filtered and washed with a little cold pyridine followed by cold ethoxyethanol. Recrystallization from ethyl acetate results in white granules.

Transition temperatures are determined in a stirred oil bath. The sealed glass tube with a small amount of crystalline cholesteryl benzoat in, is heated within a bath. The temperature of the bath is measured. The crystal melts at a melting temperature of around 150 °C into a cloudy liquid that, upon further heating, clears at the clearing temperature of between 170 °C and 180 °C.

The cholesteryl benzoat is not very appropriate for other experiments. Still, heating a few small crystals on the glass using a spirit burner allows for good observation of the phase transitions and colours of a material in a liquid crystalline phase and, at lower temperatures, of a crystalline phase between crossed polarizers under the microscope.

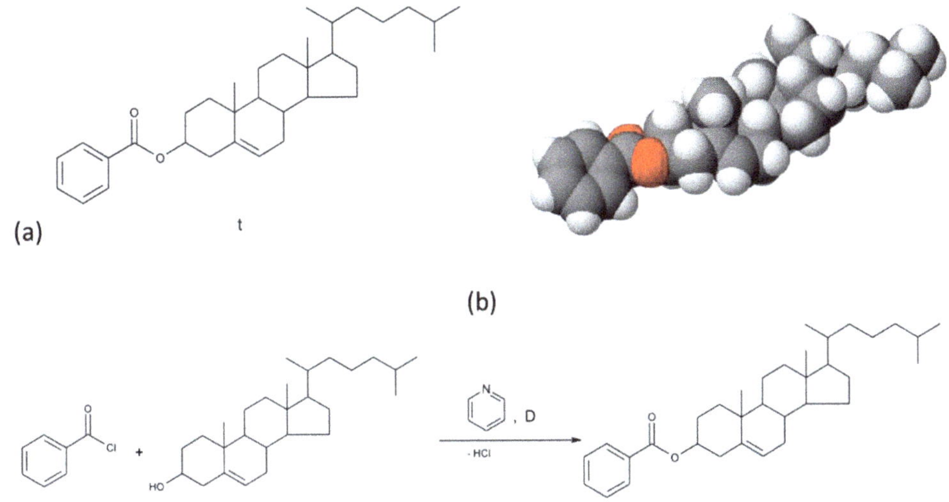


**Figure 2.2.** (*a*) The formula and a model of the cholesteryl benzoat molecule. (*b*) The synthesis of MBBA consists of these steps, given by chemical formulae.

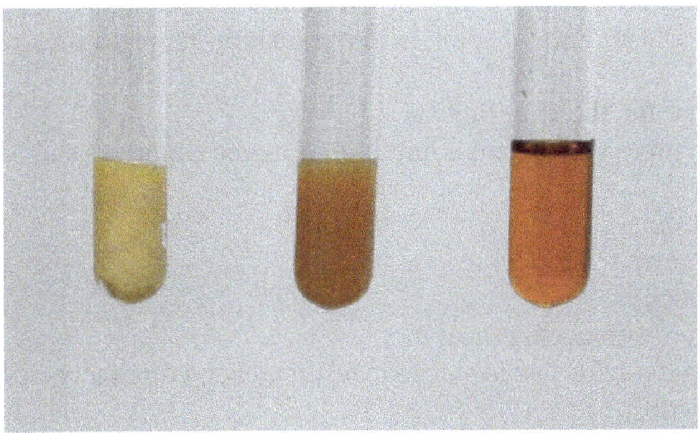

**Figure 2.3.** (left) At temperatures below 20 °C the MBBA is a yellow sugar-like crystal. (middle) At temperatures between 20 °C and 40 °C the MBBA is an opaque liquid. (right) Above 40 °C the MBBA is a yellow-brown transparent liquid.

## 2.2 Transition temperatures

The experimental measurement of transition temperatures is a very simple method to verify the synthesis. As already mentioned in the subsection 2.1, some liquid crystals are simple to synthesize and are also appropriate for school experiments, allowing students to learn about the basic concepts of liquid crystals and their special properties. They have one liquid crystalline phase, the nematic or chiral nematic phase only. The nematic phase is visually recognized as an opaque and rather viscous fluid that appears between a clear isotropic liquid and a white or yellow powder-like or sugar-like crystal (figure 2.3). This experiment (described below) is very attractive as students can visually observe phases and the phase transitions between them. It introduces a new phenomenon—a new phase—to students in a very simple way.

**Experimental aims**

- To observe changes in visual appearance during the phase transitions.
- To realize that the same visual appearance in a temperature interval is an indication of the existence of a phase.
- To realize that densities of phases in the same material differ and that a density of a crystal is higher than a density of the liquid crystalline phase, and the same is true for the liquid crystalline phase versus the isotropic liquid.
- To measure the transition temperature from the crystalline phase to the nematic liquid crystalline phase.
- To measure the transition temperature from the nematic liquid crystalline phase to the isotropic liquid phase.
- To test the quality of synthesized liquid crystal by comparing the reported transition temperatures in catalogues and the measured ones (if students have synthesized a liquid crystal by themselves).

The first three aims are of an observational nature. The last three are measurements of specific properties of liquid crystals, the transition temperatures.

**What is required for the experiments**

- Two test tubes with a small amount of frozen liquid crystal MBBA.
- Hairdryer.
- Water bath.
- Thermometer.
- Mixing stick.
- Hot plate or immersion heater.
- An alternative: for less accurate measurements, containers with hot and cold water satisfy.

The procedure is appropriate for a liquid crystal with transition temperatures between zero and up to 50 °C, like MBBA or 5CB. For materials with higher transition temperatures the procedure is essentially the same, however, higher temperatures need more care and the observational part is not so easy.

Test tubes filled with a liquid crystal are immersed in water mixed with ice cubes until they crystallize, that is the liquid crystal becomes solid. Test tubes with a frozen liquid crystal should be prepared in advance.

Each group should be given two test tubes of the same material. Both test tubes, containing a liquid crystal in its crystalline state, should remain immersed in icy water. The first test tube is used for the first part of the experiment, which is essentially observation, and the second is used for the measurement of transition temperatures.

As students are usually not familiar with liquid crystals, the observational part of this experiment provides their first experience with them. Therefore we suggest that, although it does not provide numerical measurements, it is advisable that students are allowed to 'play' with this exotic substance.

Students first warm the sample by hand and observe the melting process from the crystalline to the liquid crystalline state. Due to the small amount of liquid crystal, the content of the tube melts rather quickly to an opaque liquid. Next, the melted liquid crystal is heated with a hair dryer. The transition to the isotropic liquid is clearly seen. Now the liquid crystal is cooled by hand or by immersing the tube into the cold water. The liquid crystalline state reappears. Heating with a hair dryer is appropriate for observations only.

As an alternative to a hair dryer, the test tube can be immersed in a glass of 'hot' water. Hot tap water is satisfactory as transition temperatures of materials appropriate for these experiments should be around 40 °C. The tube content melts into the clear liquid. Removing the tube from the hot water, the liquid crystal returns to the liquid crystalline state shortly after.

The transition to the isotropic phase should be observed several times as it is clearly visible and fast. The changes of enthalpy at the transition to the crystalline phase are much larger and the process of crystallization is not quick enough to allow for several observations of crystallization by simply immersing the test tube into the icy water. Nevertheless, it is worth to try at least once, although for very pure samples or samples several years old, crystallization at zero degrees is not always successful.

At this point, the lecturer introduces the two transition temperatures: the *melting temperature*, at which the phase transition from the crystalline to the liquid crystalline phase occurs, and the *clearing temperature*, at which the transition from an opaque liquid crystalline state to the transparent isotropic liquid occurs. Even if students have had preliminary lectures, where the names for transition temperatures were discussed, one should remind students of the names at this point. Discussion with students rather than lecturing is favoured during practical work.

The second point that needs reflection is the concept of 'phase' as a state of material that has certain specific properties that exist over a wider temperature range. At this point, differences between the three phases are evident: the crystalline phase does not flow, the liquid crystalline phase flows and it is opaque, and the isotropic liquid flows and it is transparent. The opaqueness is the property of the phase, when the liquid crystal is in bulk. The opaqueness does not disappear if the sample is left at temperature between the melting and the clearing temperature, like happens to the mixture of crushed ice and warm water.

The third observation that could be stressed if the lecturer believes that it could be of further use, or that it is conceptually interesting, are phase densities. The common experience of students is that crystals have a smaller density than liquids in the same materials, as their experiences with transitions of a melting type are mostly limited to water. The melting and the clearing processes in the liquid crystal clearly show it is not uncommon for the lower temperature phases to have larger densities. The lecturer should mention that the water is actually an exception and that for most materials densities of phases at lower temperatures are higher than densities of phases at lower temperatures.

By playing with the test tubes containing liquid crystals, students have gained some personal experiences that have allowed them to construct the conceptual network linking various points related to liquid crystals. They are prepared for quantitative work and the more detailed measurements of liquid crystalline properties.

The container with the mixture of crushed ice and water, in which the second test tube with a liquid crystal in is immersed, is placed on the hot plate. The remaining ice is removed and the thermometer is put in the water. The water is slowly heated and constantly mixed using a mixing stick. As the crystal begins to melt, the melting temperature is recorded as the temperature of the water bath. After a while, the liquid crystal begins to clear and the clearing temperature is recorded in the same way.

As an alternative, one can measure the melting and the clearing temperatures by simply adding warmer water to the cooler water and measuring the temperature of the mixture. Using the alternative method, one can play around with transition temperatures, going through the transition temperatures, up and down several times.

Both measured temperatures can serve as a test for the quality of the synthesis. If temperatures are close (to a few degrees) to the reported temperatures [1], it means that the synthesized sample has been purified well and the synthesis was successful. The liquid crystal can be used for experiments described later.

If the sample is not very pure, the transition temperatures are lower. Sometimes liquid crystals do not crystallize in icy water or the temperature range of the liquid crystalline state is rather narrow. If an opaque state does not appear at all, the synthesis was not successful.

Another aspect can be considered if several groups of students measure transition temperatures. If all the students use the same material, for example in sealed test tubes, the statistics of the measured transition temperatures can be discussed. If students measure the temperatures of the liquid crystals synthesized by themselves, they can compare the quality of the syntheses, compare their lab notes and try to establish the reasons why one synthesis was more successful than another.

*Tricks of the trade.* The test tube should be small enough that the melting and clearing processes in the liquid crystal immersed in the water is relatively quick. Still, the method is not very accurate and the accuracy of the measurement is only to within a few degrees. Also, an alternating warm–cold water mixing method is not very accurate. However, the main purpose of the experiment is not the exact determination of transition temperatures but more to familiarize students with the concept of phase transition. For accurate measurements other professional methods and devices have been developed.

*Suggestions for the lecturer.* If other types of liquid crystals are available, allow for measurements of one other type of liquid crystal as well. If time allows, let groups measure transition temperatures for at least two types of liquid crystals. If not, different groups can measure different crystals. This allows students to recognize the transition temperature as a material property.

**Additional tasks**

1. Imagine the following experiment. Material A is inside a test tube and looks powder-like when immersed in a mixture of water and ice. If a vial of material

A is immersed in boiling water, it is opaque. Material B is an opaque, very viscous liquid when in a mixture of water and ice, and is a clear transparent liquid in boiling water. What can one tell about the melting and clearing temperatures for these two materials? Compare the melting temperatures for both materials. Compare the clearing temperatures for both materials.
2. Describe the differences in observations at the melting and clearing temperatures for a fictitious liquid crystal, where density increases from one phase to another upon increasing temperature.

## 2.3 Ordering of liquid crystals

Liquid crystals are anisotropic fluids, but their anisotropic properties are recognized only if they are ordered. What does 'ordered' actually mean? It seems like an oxymoron as we have already discussed the orientational order as an essential property of liquid crystals. Liquid crystals in larger amounts, in drops or in containers like tubes, are seen as an opaque liquid. The orientational order exists, but the order is local in the sense that a correlation length for orientational order is a few hundreds of micrometers. The sample consists of groups of molecules called domains with a diameter of approximate correlation length containing approximately $10^{15}$ molecules. Molecular groups differ from one another in the direction of a director but not in the 'quality' of the order in general. The director direction changes rather abruptly from one domain to another, forming a defect in a director field called a domain wall. Within a domain wall, the director changes and also the order itself is usually lower. As the width of a domain wall is shorter than the wavelengths of visible light, and optical properties of domains strongly depend on a director orientation, light reflects and refracts on the domain wall. As domain walls are oriented in various directions, light also reflects and refracts in different directions and intense scattering occurs due to domain walls. The liquid crystal scatters the light in all directions and is therefore seen as an opaque liquid (see figure 2.4).

In order to study optical properties of liquid crystals and to benefit from their properties in applications, the liquid crystal has to be ordered macroscopically. The director has to have the same orientation over the whole sample. The first approach to ordering is to take a sample with a very small volume consisting of one domain only. As such a small sample cannot be used in devices, a trick is used—the size of the domain is limited below the correlation length in one dimension only. The liquid crystal is put in a container that is narrower than a few hundreds of microns in one dimension. Such containers are called cells and are in principle very simple. Cells consist of two parallel glasses separated by spacers that determine the thickness of the cell. The thickness of cells in devices is a few microns. For laboratory research purposes thicknesses vary from cells without spacers, leaving the cell thickness in submicron regions, up to several hundreds of micrometers, approaching correlation lengths, depending on the research goal.

As the optical properties of a device depend on a liquid crystal structure in a cell, to prepare cells for applications, the orientation of long molecular axes is controlled. Several methods exist that determine the orientation of long molecular axes with respect to the glass normal. For research studies two typical orientations are known. If the

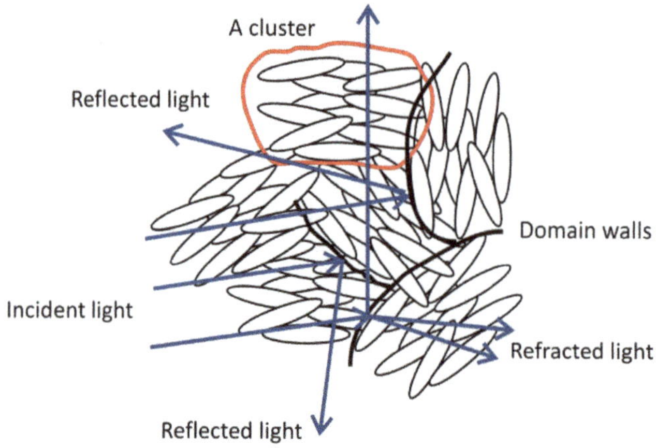


**Figure 2.4.** Average direction of molecular long axes differ from one domain to another. Light is reflected and doubly refracted at domain walls resulting in scattering of the light and an opaque milky appearance of a liquid crystal in bulk. Note that domains in the schematic diagram are symbolic. Real domains contain significantly larger numbers of molecules then indicated in the schematic diagram.

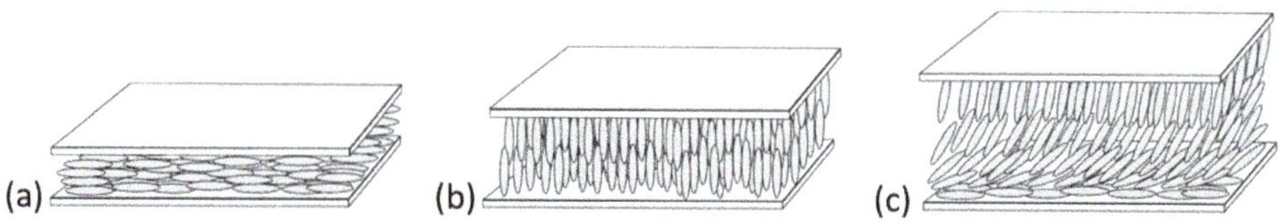


**Figure 2.5.** (*a*) A schematic of a planar cell. (*b*) A schematic of a homeotropic cell. (*c*) A schematic of a hybrid cell with a combination of planar and homeotropic treatments on the surfaces.

director is parallel to the glass surface, the cell is called the *planar cell* (figure 2.5(*a*)). If the director is parallel to the layer normal, that is perpendicular to the glass surface, the cell is called the *homeotropic cell* (figure 2.5(*b*)). For several studies, special cells have been designed where a treatment of the surfaces imposes the structure of the cell in various ways. For such cases surfaces have different properties with respect to favourable orientation and cells are often called *hybrid cells* (figure 2.5(*c*)).

Below we present a simple hands-on fabrication of such cells, and the conditions which have to be fulfilled for the cell to be considered satisfactory for teaching requirements. We also give few ideas of what could be studied with such cells.

### 2.3.1 Manufacturing the cell

Cells that are manufactured for research purposes and for later use in devices are made using very rigorous treatments. They are assembled in a clean, dust free room under strictly controlled conditions. Laboratories often have a clean room for assembling cells according to the specific needs for experiments or they buy cells from providers. If experiments do not require extreme quality of order, the cells can be assembled under lesser control. For school laboratory purposes satisfactory cells can be manufactured by very simple means.

**Experimental aims**

- To learn how to manufacture a cell with non-treated surfaces.
- To become familiar with the appearance of cells with various thicknesses between crossed polarizers.
- To observe visual changes in a liquid crystal during heating and cooling.
- To observe visual changes in a liquid crystal under weak external pressure.
- To recognize a main difference between an optically isotropic and an anisotropic material and its relation to the phase transition.

**What is required for the experiments**

- Liquid crystal that is in a liquid crystalline state at room temperature.
- Object slides usually used for regular school microscopy.
- Cover slides for the same purpose as above.
- Spacers: adhesive tape, food wrapping foil, transparency, cover slide.
- Two polarizers.
- Overhead projector.
- (or) a supported piece of matt glass illuminated by a white light from below.
- Microscope.
- Hair dryer.
- Alcohol and a piece of clean cloth (if new object slides are not available).
- A dripping-tube.

The equipment needed for the experiment can be seen in figure 2.6. A new object slide is taken from the box or an old one is cleaned by alcohol as much as possible. Equal spacers are put at a distance that is few millimetres shorter than the dimension of the cover slide. For very thin cells, spacers are not needed. For cells with a thickness of a few tenths of a micrometre, one, two or more layers of a thin food wrapping foil are used. Even thicker cells are obtained by a few layers of adhesive tape. For a thickness of

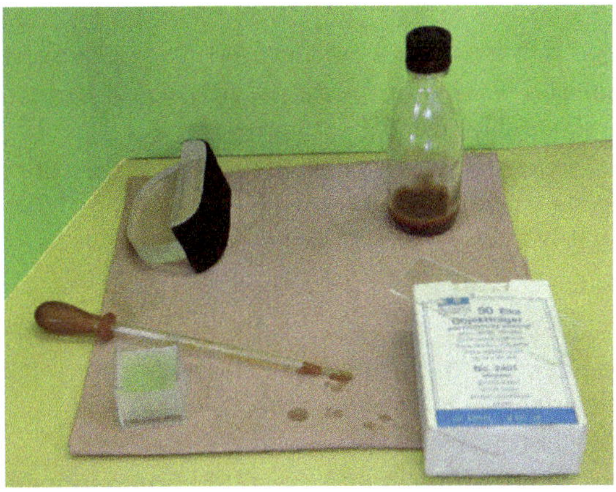

**Figure 2.6.** Object and cover glasses, a liquid crystal, a holder with velvet, and a dripping tube (Photo: Maja Pečar).

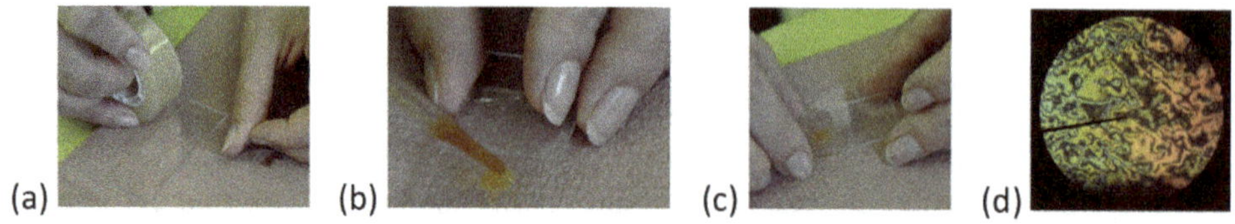

**Figure 2.7.** (*a*) Tapes are positioned as spacers. (*b*) A drop of liquid crystal is put between spacers. (*c*) The cover glass is positioned onto a drop and both spacers. (*d*) The final cell between the crossed polarizers through a microscope (Photo: Maja Pečar).

up to 400 micrometers, which is already close to the correlation length for the orientational order, a few layers of transparencies or two cover glasses are used.

A very small drop of a liquid crystal is placed between the two spacers and the cover slide is laid over a drop, forming a bridge between two spacers. Liquid crystal spreads over the area between the slides easily. The cell is prepared for observation. The step-by-step assembly of the liquid crystalline cell is illustrated in figure 2.7.

The cell is placed between crossed polarizers. If the liquid crystal is in the liquid crystalline state, the light is transmitted through the sandwich consisting of the cell containing a liquid crystal between the two polarizers. The appearance depends on the type of liquid crystal, but above all, it depends on the thickness of the cell. Thinner cells between crossed polarizers are only weakly transparent, while cells with a thickness of a few to a few tenths of micrometres reveal colours that change with temperature. A cell with no surface treatment also reveals several defects that are characteristic for various phases found in liquid crystals [4].

What to do with such a cell? As has already been said, cells of different thicknesses have various appearances. It is advised that different groups assemble cells of different thicknesses. Students should observe the cell for a long time, as a liquid crystal flows and the appearance changes as a liquid crystal slowly orders. Cells are than exchanged and each student should study each type of cell under a microscope. Cells can also be observed directly or on an overhead projector but several details cannot be seen.

*Tricks of the trade.* All materials that are used for spacers, except cover glasses, are anisotropic. The food wrapping foil or an adhesive tape reveals several optical phenomena very similar to liquid crystals. In order to focus on liquid crystals only, a non-transparent adhesive tape (like a medical tape or a piece of paper) should cover the parts with the spacers to block the transmission of light. The optics of tapes and transparencies is studied in more detail in chapter 4.

**Additional tasks**

1. Measure the thickness of various spacers. Calculate the thickness from the length of new adhesive tape or a roll of food wrapping foil, or measure the thickness of several transparencies at the same time.
2. Observe a cell filled with a liquid crystal between crossed polarizers. Draw defects (dark lines) and try to sort them. What are the characteristics of the observed defects?

3. Observe a cell between the crossed polarizers. Observe the colours. Are any colours usually found in the rainbow missing? If so, which? What are the sequences of colours at various places? Which colours are found close to each other?

   Rotate one of the polarizers by 90° to make the polarizers parallel. Compare the colours with the colours observed for the same cell through crossed polarizers.
4. Observe a very thin cell through crossed polarizers. Tilt the cell in order to observe the light that is transmitted through the cell in an oblique direction. How does the appearance of a tilted cell differ from the cell that is perpendicular to the incident light?

### 2.3.2 Planar cell

Observations of a cell between crossed or parallel polarizers where glass surfaces have not been specially treated reveal several features. Several black lines are found that indicate defects. Intensive changes of colours are observed. All observable changes appear due to the variation in director and due to variation in the order parameter, which is lower in defects. With many variations in the director and the order, the cell cannot be used in research studies or applications. Therefore the variation in director through the cell has to be eliminated. The simplest procedure for this purpose is simply rubbing it with a hairy piece of cloth such as velvet. Rubbing in one direction causes tiny scratches on the glass surface. As elongated liquid crystalline molecules prefer orientation that maximizes the contact area of the molecule with the glass, they orient along the scratches. Although only molecules at the surface are affected by interactions with glass and the average orientation of long molecular axes is preferred in the direction of the scratches, the orientation of the director is transferred to the interior of the cell. The director in the cell is not arbitrary any more and molecules in the whole volume of the cell are oriented in the rubbing direction. This orientation is called planar, as discussed in the introduction of this section.

**Experimental aims**

- To learn how to manufacture a cell with planar alignment.
- To compare the visual appearance of the planar and the non-treated cell.
- To assess the quality of ordering.
- To find a director profile experimentally.

**What is required for the experiments**

In addition to the equipment needed for assembling a cell with non-treated surfaces, one needs a clean piece of hairy cloth like velvet or a piece of synthetic carpet. The cloth is simplest to use if it is glued onto a holder (figure 2.8).

An object slide is cleaned with alcohol as much as possible, even if it is new from the box. The velvet is cleaned by alcohol too. The object slide is rubbed in a predefined direction. The same procedure is used for the cover slide.

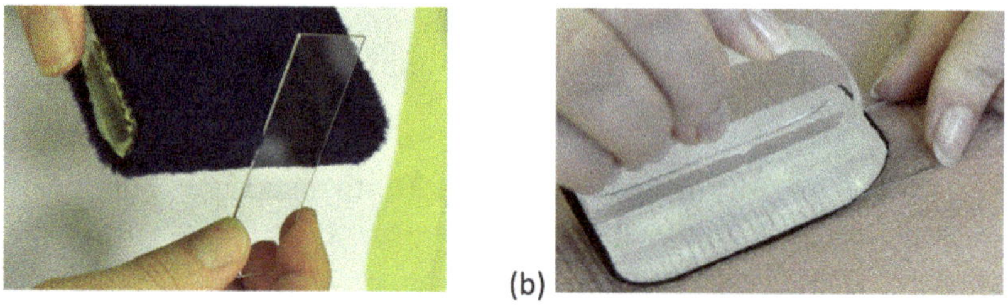

(a) (b)

**Figure 2.8.** (*a*) The cloth glued around a piece of plexi glass. (*b*) Specially designed holder that allows for more controlled pressure and direction of rubbing.

The assembly procedure is the same as for the cell with untreated surfaces. Equal spacers are put in at a distance that is few millimetres shorter than the dimension of a cover glass. For very thin cells, spacers are not needed. For cells with a thickness of a few tenths of micrometres, one, two or more layers of thin food wrapping foil are used. Even thicker cells are obtained by using a few layers of adhesive tape. For thicknesses of up to 400 micrometers, which is already close to the correlation length for the orientational order, a few transparencies or simply two cover glasses are used.

The object glass is preheated using a hair dryer. A very small drop of a liquid crystal is put between two spacers and the drop is quickly covered by the cover glass. The best orientation is obtained if the space between the object and the cover glass is filled by a liquid crystal in the isotropic phase. When the cell is cooled the molecules align along the scratches more easily than if the material is drawn into the cell by capillary action in the liquid crystalline phase. The cell is prepared for observation. The cell should look homogeneous (of mostly the same colour, with only a few defects). The alignment can be improved if the cell is again heated to above the isotropic transition and then cooled down again.

One should be aware that molecules align along the rubbing direction, therefore the director will be parallel to the rubbing direction. If a specific direction of director is required, for example for the preparation of wedge cells, the rubbing direction is determined. If not, students can choose any rubbing direction providing the parallelism of rubbing at the upper and lower slide surface is ensured.

The cell is placed between crossed polarizers. The cell is rotated between crossed polarizers and the transmission orientations of the polarizer closer to the observer, for which the cell appears (mostly) dark, is marked on the cell surface. What is the relation of these two directions with respect to the rubbing direction?

Next, the rubbing direction should form an angle of approximately 45° with the transmission directions of the polarizers. If the liquid crystal is in the liquid crystalline state, the light is transmitted through the sandwich consisting of the cell containing a liquid crystal between the two polarizers. If the alignment is good, the transmission of light is almost homogeneous, and for cells, where colours appear, the colour is more or less uniform. The effects of rubbing are usually seen on ordering. Defects could still be observed but they usually look different to those in the non-treated cells.

After assembling the cell, students observe the cell under crossed and parallel polarizers. As already mentioned, cells of different thicknesses have various appearances. Each group of students should assemble a cell of a different thickness. Cells are then exchanged and each group can study each type of cell under the microscope. Cells can also be studied by placing them on a computer tablet or on an overhead projector.

Although the optical properties of ordered liquid crystals in the cell will be discussed later, a short explanation can be given at this point too. The non-polarized light splits into two mutually perpendicularly polarized beams in the liquid crystal. One beam is polarized parallel to the director, the other one is polarized perpendicular to it. If the first polarizer that polarizes the incident light is parallel or perpendicular to the director i.e. to the rubbing direction, only one beam is present in the liquid crystals and the transmission through the liquid crystal does not change its polarization properties. Therefore it is absorbed by the second polarizer with the transmission direction perpendicular to the first one. By this simple experiment the direction of the director is determined as one of the two perpendicular directions. For final determination of which of the two directions is the director, more experiments are needed.

If the polarization of the incident beam forms a general angle with a director, the beam splits into two beams. After passing the liquid crystals, beams are phase shifted and the shift depends on the optical path, i.e. on the wavelength of light. So the elliptical state depends on the colour of light. The second polarizer therefore absorbs different wavelengths differently and the transmitted light becomes coloured. If a director in the cell is constant, as well as the quality of the order and the thickness, the colour of the transmitted light is the same everywhere. For cell assembled by hand this is usually not true and they vary in colour.

*Tricks of the trade.* The influences of the anisotropic properties of spacers should be avoided, as they may distract students from the observations using the non-transparent cover, as before.

In well-equipped laboratories, professional polarizing microscopes are usually available (figure 1.1). Although such microscopes provide large magnifications and easy rotations of polarizers, we suggest that at least one cell is studied by the simple means described above. Enough time should be allowed for students to play with the cell inserted between two polarizers. Various orientations of cells with respect to one or other of the two polarizers should be studied and discussed. These activities provide a link between the purpose of the polarizing microscope and the meaning of the polarizers' manipulation. When students have had experience with the simple set-up, they easily proceed to professional equipment.

*Suggestions for the lecturer.* Several experiments can be done with such a simple cell. The quality of the alignment is verified under a simple microscope available for biology lessons in school. If the quality of alignment is relatively good, in some areas of the cell at least, the cell can be used for optical experiments, for example, for studying states of elliptical polarization.

**Additional tasks**

1. Compare the appearance of cells of different thicknesses between crossed polarizers. At which thickness do colours start to appear? At which thickness are colours limited to green and magenta? At which thickness is the transmitted light grey again?

2. Choose a cell that exhibits colours. Insert the cell between crossed polarizers with the transmission direction of the upper polarizer at 45° with respect to the rubbing direction. Rotate the upper polarizer and observe the colours. Describe the changes in a colour.
3. Place the cell that exhibits colours between crossed polarizers. Choose an angle between the transmission direction of the upper polarizer and the rubbing direction where colours are most clear. Heat the cell with a hair dryer. What happens? The most detail can be observed if the cell between crossed polarizers is observed through a microscope.

   Switch off the hair dryer. Observe the cell during cooling. Be patient. The experiment lasts for few minutes at least.

### 2.3.3 Homeotropic cell

In previous subsections 2.3.1 and 2.3.2 students learned how to assemble a cell. In this subsection we consider how to assemble cells where molecules have their long molecular axes oriented perpendicularly to the surfaces, as in figure 2.5(*b*). Such alignment is called homeotropic. Such orientation of molecules in the cell is used for various experimental studies of liquid crystals. The treatment is used for studies of defects, for observations of conoscopic figures and for studies of other properties.

**Experimental aims**

- To learn how to manufacture a cell so that molecules are aligned homeotropically.
- To compare the visual appearance of homeotropic, planar and non-treated cells.
- To assess the quality of ordering.
- To compare defects in a homeotropically aligned cell and in cells with planar alignment.

**What is required for the experiments**

In addition to the material that is needed for assembling the non-treated cell, the following is needed:

- Detergent for dish washing or a liquid soap; also lecithin can be used but the achieved alignment is not significantly different.
- Distilled water.
- A beaker and stands for drying the glasses.

Prepare a solution of a detergent or a soap and a distilled water. Clear detergent or liquid soap should be used. For approximately 200 ml of solution, one teaspoon of detergent is added and mixed, and the solution is left for bubbles to disappear.

Clean the object glass with alcohol and leave it for a minute or two to dry. A clean object glass is submerged into the solution and left for a few minutes, then removed from the solution. The glass is left on the stander to dry. Before glasses are used they have to be completely dry. The same procedure is repeated for the cover glass.

The assembly procedure is the same as for the cell with untreated surfaces. Also here the preliminary heating of both glasses above the clearing temperature allows for better alignment of the liquid crystal. After the cell is assembled, it should be left without shocks for a few hours or a whole day for the order to establish. Even later, it is better that the cell is not moved around much as the order is rather sensitive to any sudden mechanical shocks.

The quality of the alignment is tested by polarizers. As molecules are oriented perpendicularly to the glass surface, the system is uniaxial and the optical axis is perpendicular to the glass surface. Therefore the two speeds of light is equal for any polarization and the phase difference does not appear. The well-aligned cell between crossed polarizers is black. However, if the incident light is oblique, the beams again split into two, the one with the polarization having a component parallel to the director and the one with polarization being perpendicular to it. The birefringence depends on the incident angle. For oblique incidence the transmitted light is elliptically polarized and the cell becomes a little transparent.

After assembling the cell, students observe the cell under crossed and parallel polarizers. Between the crossed polarizers the well aligned cell is black. The cell observed under an angle is not completely dark. As already mentioned, cells of different thicknesses have various appearances. Each group of students should assemble a cell of a different thickness. Cells should then be exchanged and each group can study each type of a cell under a microscope. Thicker cells have usually more defects.

*Tricks of the trade.* Homeotropic alignment is more stable if the liquid crystal is in the smectic phase. If 5CB is available, which is at room temperature in the smectic phase, the order is quite stable and not so much care to avoid shocks is needed. The experiments suggested yield very similar results for homeotropically aligned liquid crystal in the nematic or in the smectic phases.

**Additional tasks**

1. Observe a thick cell filled with a liquid crystal between the crossed polarizers. Is it possible to observe defects? How do they look compared to the cell with planar alignment?
2. Observe a thick cell between the crossed polarizers. Tilt the cell. How does the appearance differ if the cell is observed directly or if it is tilted with respect to the observation direction?
3. Observe a very thin cell between the crossed polarizers. Can you observe defects? Compare the defects with the defects observed in a thick cell.

### 2.3.4 Cells with different surfaces

The cells that were assembled in previous sections are usually used for research purposes. Studies of defects reveal the phases of liquid crystal. Textures, that is an appearance of the liquid crystal in the cell between crossed polarizers, change with temperature and indicate different phases and temperatures of the phase transitions. Combinations of texture observations of planar and homeotropic cells give hints for structures of the phases at the microscopic level. Additional information is obtained by studies of textures in external fields. While studies of magnetic field effects do require a controlled alignment only, studies of structures in an electric field require glasses covered by conductive layers. The latter are usually not easily available, but laboratories studying liquid crystals are usually willing to provide a slide with conductive layers. The procedure for assembling cells from glasses with conductive layers is the same as discussed in the previous three subsections. Therefore the assembling process will not be discussed in detail.

More detailed studies of liquid crystal require several tricks that allow the study of specific properties. For example, if one wants to study the properties of a liquid crystal associated to the bent of a director, a hybrid cell is required, where the director of a liquid crystal in a large part of a volume is bent. This is achieved in cells where one of the surfaces is treated for a planar alignment and the other for homeotropic (see figure 2.5(*c*)). The basis of LCD function is the cell where the twist is important. This is obtained by surfaces favouring planar alignment that are rubbed in different directions.

In this subsection two different cells are discussed, the hybrid and the twisted cell. Both are meant more for information for students, and for the possibility of observing the phenomena, than for serious studies. The latter are left to research laboratories. It is not difficult to assemble a hybrid or twisted cell if students have previous experience with planar and homeotropic alignment. Therefore it would be a pity not to use the opportunity for further studies of cell properties which may be extended to serious projects.

**Experimental aims**

- To learn how to assemble hybrid and twisted cells.
- To compare the visual appearance of a hybrid or twisted cell with homeotropic cells, planar cells and cells without treated surfaces.
- To observe defects typical for various cells.
- To study the transmission of light through cells with unequally treated surfaces.

**What is required for the experiments**

No special equipment is required besides the equipment needed for assembling cells already discussed. Slides required for a planar or homeotropic alignment are prepared according to the procedures already discussed in subsections 2.3.2 and 2.3.3.

*The bent hybrid cell.* Take two object slides. Prepare one object slide as for a planar alignment. Rub the slide surface along the longer edge. Prepare the second object slide as for the homeotropic alignment. Add spacers of the thickness desired for a study, place a drop of a liquid crystal on the object slides with spacers. Heat both object slides with a hair dryer and place the second slide on the spacers when the liquid crystal become isotropic i.e. its temperature and the temperature of slides is higher then the clearing temperature.

*The twisted cell.* The assembling procedure is the same as for the planar cell, except that the rubbing direction of the two slide surfaces is not parallel. To obtain a uniform twist, the angle between the rubbing directions of the slide above and below should not be equal to 90° if the liquid crystal is racemic. If the liquid crystal is chiral or a racemate is doped by a chiral dopant, this question need not be considered.

*The wedge cell with a non-treated surface.* The object slide is prepared as for the planar alignment. The spacer is added at one side only. The thickness of the spacer should be large, for example a few cover slides. A drop of a liquid crystal is put on close to the spacer. The cover glass is placed in such a way that on one side the glass is in direct contact with an object slide, on the other side it rests on the spacer.

For all three cells, the methods for taking observations are essentially the same. The cells are placed between the crossed polarizers and are observed under the microscope. Additional rotations of cells between crossed polarizers or rotations of one of the polarizers provide additional interesting observations. Cells with different thicknesses allow the study of a series of phenomena such as the transmission of light, defects, domains of different twists and similar.

The effects of thickness can be studied in a single cell in the wedge cell. A cell with one non-treated slide provides information about the correlation length for a director order along the short molecular axes.

**Additional tasks**

1. Observe the hybrid cell between crossed polarizers. Compare the defects observed in a hybrid cell to the defects observed in homeotropic or planar cells.
2. Study hybrid cells of different thicknesses between crossed polarizers under a microscope. Study the defects.
3. Observe the twisted cell between crossed polarizers under a microscope. Draw the defects. Compare the defects to the defects observed in twisted cells with different thicknesses.
4. Observe a twisted cell under crossed polarizers. Find the orientation of the second polarizer for which the transmitted light is minimal.
5. Study cells with different relative orientations of rubbing directions on slide surfaces.
6. Observe the wedge cell under crossed polarizers. Find the thickness, for which the sample is not ordered through the whole bulk. This thickness gives an estimation of the correlation length for the nematic order perpendicular to the director.

## 2.4 Optical anisotropy of liquid crystals

Liquid crystals are strongly anisotropic materials in the region of optical wavelengths. Phases like nematic or smectic A are uniaxial with refractive index values of about 1.8 for polarization parallel to the director and about 1.5 perpendicular to it, with corresponding birefringence of around 0.3. Optical anisotropy is therefore extremely high.

The large difference in refractive indices usually results in a large phase difference between partial beams with two principal polarizations and the consequent transmission of light through the cell between crossed polarizers. The colours of transmitted light give us information about the phase difference leading to the difference in refractive indices, and consequently allow us to deduce the structure of new phases. However, with easy manipulation of the liquid crystal structure within a cell by wisely planned treatment of cell surfaces and external fields, the optical properties of the cell are manipulated and allow them to be applied; they are most widely used in screens.

Due to the fast rotation of molecules around their long molecular axes, the nematic phase is always uniaxial. Experiments with uniaxial anisotropic materials are analysed more easily than the same experiments performed with biaxial anisotropic materials. However, one should note that more advanced problems involve biaxial materials, as several liquid crystals have structures that are strongly biaxial, at least locally.

Experiments with nematic liquid crystals in cells provide several experiences for students. They allow the study of general optics phenomena in uniaxial anisotropic materials, that is, the effects of transparent anisotropic material on the polarized light. However, to introduce the various polarization states of transmitted light or vice versa, to deduce properties of anisotropic material from the state of transmitted light, is not a simple quest. In theoretical treatment the lecturer usually assumes known directions of polarization for the two partial waves in the anisotropic material. These directions are found by a simple test in uniaxial materials only; in biaxial materials finding these directions is not straightforward [5], which may lead to false conclusions in more demanding studies.

At the beginning of each subsection the necessary theoretical background is briefly given. This part is not meant for the study of optics as there are several perfect textbooks [6, 7] covering the topic. The theoretical background is meant for establishing the link between observations, measurements and theoretical understanding of the phenomena studied. Although the theoretical part is at the beginning, it is suggested that students do some preliminary observations before the lecturer presents equations and derivations.

### 2.4.1 Linear polarization

Light is an electromagnetic wave. A light from a usual light source is not polarized, that is, when observing an electric (or magnetic) field at a specific point in space, the direction of the electric field changes in time and cannot be predicted in advance. Such light is called *non-polarized* light. When the electric field changes in a predictable way with time, the light is called *polarized.* There are several types of light polarization, also called polarization states. If the electric field in the electromagnetic wave oscillates in a constant plane, the light is *linearly* polarized. Linear polarization is considered to be the most simple of various polarization states. How does the light become linearly polarized? There exist several light sources that already emit polarized light, for example, lasers and sometimes also laser pointers. In general, the light becomes polarized after transmission through a polarizer. The polarizer absorbs the light associated with one component of the electric field and the transmitted light is polarized perpendicularly to this direction. The transmitted light therefore has the well defined direction of an oscillating electric field, which is called a *polarization direction.*

Modern polarizers are plastics, where long polymeric molecules are partially straightened and ordered. The component of the electric field in the electromagnetic wave along the polymeric molecules induces micro currents and the energy associated with this component of the electric field is absorbed. The transmitted light is polarized perpendicular to this direction. The direction is called the *transmission direction*. If the incident light is already polarized, the same consideration is valid. The transmitted light has the component of the electric field parallel to the transmission direction only. The amplitude of the electric field of the initially non-polarized light passing through two polarizers placed one after another depends on the angle between the transmission directions of both polarizers $\alpha$ (figure 2.9(*a*)), is

$$E_t = E_0 \cos \alpha \tag{2.1}$$

and the corresponding intensity of the transmitted light is

$$I = I_0 \cos^2 \alpha \tag{2.2}$$

where subscript 0 marks the amplitude of the electric field in the incident electromagnetic wave as well as for its intensity. From equation (2.2) it is clearly recognized that when the transmission directions of the polarizers are perpendicular, no light is transmitted. The polarizers are called *crossed* (figure 2.9(*b*)). Crossed polarizers are the most common setting in the various studies of anisotropic materials.

One polarizer can be used for an analysis of the polarization state of light. It is called the *analyser* in such role. The incident light is linearly polarized if one orientation of a polarizer exists where the incident light is not transmitted. The distinction of other states of polarized light, such as a circular and an elliptical polarization from the partially polarized and non-polarized light, is not possible with only an analyser. How to distinguish between these states will be discussed later.

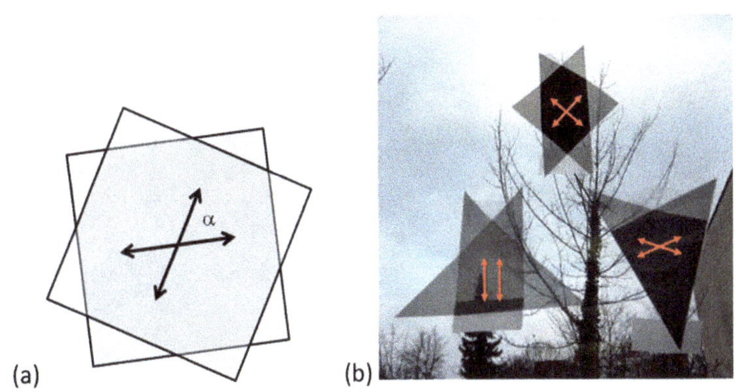


**Figure 2.9.** (*a*) Two polarizers, one on top of the other. The angle $\alpha$ is formed between the transmission directions of the polarizers. (*b*) Pairs of polarizers on the window. The transmission direction of the polarizers are marked.

The plastic, from which polarizers are made, is linearly dichroic, that is, the absorption is different for different polarizations. The absorption is characterized by a complex index of refraction. The imaginary part of the refractive index is related to the absorption. The quality of the polarizer depends on the ratio of the absorption indices for the two polarizations and on the thickness of the polarizer. As the following experiments do not require extreme accuracy, any polarizers are appropriate. A simple test consists of shining a laser pointer beam on crossed polarizers. If the transmitted light cannot be observed, the quality of polarizers is more than satisfactory.

**Experimental aims**

- To become familiar with the function of polarizers.
- To find the transmission direction of an unknown polarizer.
- To recognize linearly polarized light and to determine the direction of the polarization.
- To measure the intensity of transmitted light that depends on the angle between the transmission direction.
- To compare the measurement to Malus' law.

**What is required for the experiments**

- One polarizer with a known transmission direction.
- One polarizer with an unknown transmission direction.
- A laser pointer.
- A device for measuring the intensity of light (a photoresistor, a photodiode, a luxmeter).
- Various light sources (a light reflected on a dielectric, computer screens, tablets, mobile phones).
- For more accurate measurements: An optical bench with fixed components described above; polarizers equipped with goniometers (figure 2.10(*b*)).

*Getting familiar with polarizers.* Take both polarizers and place them on a piece of white paper, one on top of the other. The lower polarizer should be the one for which

(a) (b)

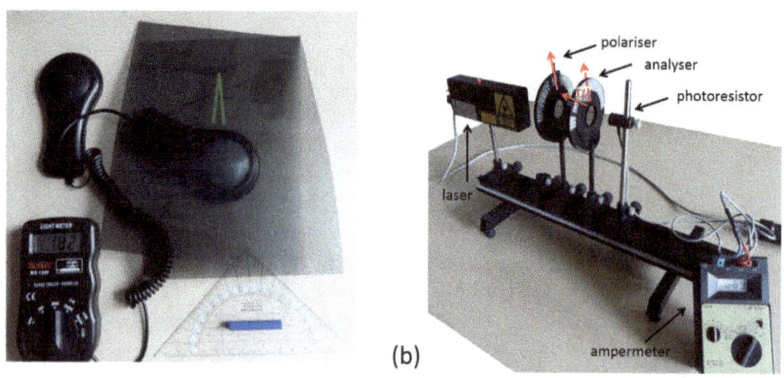


**Figure 2.10.** Requirements: (*a*) Equipment for first contact with polarizers using a simple luxmeter. (*b*) Optical bench equipped for more detailed measurements.

the transmission direction is known and marked on it. Rotate the upper polarizer slowly and observe.

Mark the transmission direction on the unknown polarizer. Use that polarizer as an analyser. Analyse several light sources: daylight, light reflected from water or from the whiteboard, light from your computer screen, your tablet or your mobile phone. The polarizer is used as an analyser. One rotates the analyser and observes the reflected or the transmitted light. The orientation of the analyser that appears dark is found only when the light is linearly polarized. The polarization direction is perpendicular to the transmission direction of the polarizer when the state is dark. If the intensity of the transmitted light changes but does not become completely dark for any orientation, the light may be partially polarized or elliptically polarized. More experiments are needed to distinguish between the two. Finally, if the intensity of light does not depend on the orientation of the analyser, the light is either circularly polarized or non-polarized. It is quite safe to conclude that if the light source is not emitted from a technical application, then the light is partially polarized or non-polarized. Polarized states are not very common in nature.

*Malus' law.* First measure the intensity of transmitted light for the parallel transmission directions of two polarizers placed on top of one another. Then measure the intensity of transmitted light for crossed polarizers. The two measurements can be used to find the proper measuring range for the device used to take measurements of the intensity of light. Next, the analyser is rotated stepwise from $\alpha = 0°$ to $\alpha = 90°$ a few degrees at a time. The intensity of light is measured. The background is estimated from the measurement for the crossed polarizers. Compare the graphical presentation to the theoretical expression for Malus' law, equation (2.2) as in figure 2.11.

Polarizers are an essential part of all screens. Almost all optical observations of liquid crystals use polarizers in one way or another. Therefore becoming familiar with polarizers is crucial for experimenting with optically anisotropic materials.

*Tricks of the trade.* Polarizers can be ordered from providers of educational equipment. They come as foils with the dimensions of normal printer paper. They can be cut into much smaller pieces. If a lecturer has access to a laboratory where polarizers

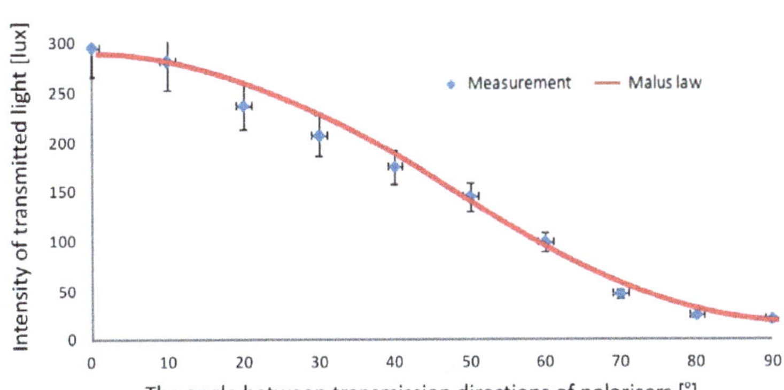


**Figure 2.11.** An example of light intensity measured through two polarizers by simple means (figure 2.10) and its comparison with Malus' law according to equation (2.2).

are regularly used, the polarizers will come in rolls. Leftovers are very common and researchers are usually happy to donate them. If this is the case, then, ideally, students would get a piece of polarizer each for later experiments outside or at home.

Plastic polarizers are not ideal and it is possible that part of the light is transmitted through the crossed polarizers. In addition, one cannot completely exclude reflections from surroundings. Therefore the background should be considered in the analysis of results.

*Suggestions for the lecturer*. There are several experiments which are stimulating and allow for several activities. The two polarizers are cut from the foil in such a way that the shape of the polarizer is rectangular but the transmission direction is 45° with respect to the sides. The transmission direction should not be marked. Such polarizers can offer students a nice challenge. One polarizer is placed on top of the other with sides parallel. Two possible situations can occur, they are either bright or dark. Rotation by 90° shows the usual Malus' law behaviour. However, if one of the polarizers is turned over, the initially dark state becomes bright or vice versa. Let students discover the explanation. Here one can also discuss the influences of previous experiences. Usually the polarizers that are used are rectangular, or sometimes square. The transmission directions are mostly parallel to the sides. This preliminary experience usually frames the way of thinking and students do not easily come up with an alternative idea, though it is very simple.

**Additional tasks**

1. Observe the bright sky above the horizon and in the zenith through an analyser. What can you say about its polarization?
2. Observe a rainbow through an analyser.
3. Observe light reflected on water through an analyser for different heights of the Sun.
4. Find other sources of partially or linearly polarized light in everyday situations.
5. Observe a 3D television through an analyser. What happens to a picture if the analyser is rotated? Observe a colleague wearing 3D glasses through the analyser. What happens to a picture if the analyser is rotated? Can you deduce the principles of 3D TV from your observations?

### 2.4.2 Double refraction

The experiments in this subsection are appropriate as demonstrations or they can be used as less demanding or more demanding laboratory experiments. They can also be extended into projects. This set of experiments is one of the most important; they need to be studied when the subject of optical properties of liquid crystals is presented, and in optics courses, when optics of anisotropic materials is studied.

When light propagates through an anisotropic material it splits into two beams with mutually perpendicular polarizations and different phase velocities. If the incident direction of light is perpendicular to the surface of an anisotropic material, the wave vector remains the same although the propagation of energy may be different. If the incident angle is not equal to zero, the wave vectors of the two beams differ and the light refracts differently according to Snell's law for each polarization. The phenomenon is called *double refraction*, as the single beam splits into two beams with different directions.

The phenomenon is usually observed in transparent anisotropic crystals, where the images observed through crystals are doubled. This experiment will be presented later, within the set of experiments illustrating accompanying concepts in chapter 4.

In liquid crystals such observation is not as straightforward, as the liquid crystal has to be ordered to reveal its anisotropic properties. The cells that guarantee the order are rather thin and the space separation of the laser beam does not occur during the transition through the anisotropic liquid crystal. Therefore another trick should be used. As refractive indices for the two polarizations are different, the propagation of light through the prism of a liquid crystal results in different directions of beams after passing the prismatic cell. The method is the same as for the prism used for decomposition of white light [8, 9] to its colour components due to dispersion. For this purpose the cells have a different thickness on each side—wedge cells are used. The wedge cells are assembled in the same way as the wedge cell discussed in subsection 2.3.4. The only difference is that surfaces have to be treated to allow for planar or homeotropic alignment of nematic liquid crystal.

**Experimental aims**

- To show an existence of two waves with different refractive indices in a planar nematic liquid crystal.
- To verify the polarization of both beams.
- To observe changes at the phase transition.
- To measure both refractive indices.
- To relate the direction of the director, the polarization of the beam and the refractive indices with respect to their magnitudes.
- To study the angular dependence of both refractive indices for a homeotropically oriented wedge cell.
- To measure both refractive indices dependent on the beam direction with respect to the optical axis.

**What is required for the experiments**

- A wedge cell with a director parallel to the wedge.
- A laser pointer.
- A polarizer.
- A ruler for measuring distances between light spots on the distant screen.
- A wedge cell with homeotropic alignment of a director.
- A wedge cell with a planar aligned director perpendicular to the wedge.
- Two prisms with an apex angle of 90° and two prisms with an apex angle of 60°, all of them with the same refractive index of close to 1.5.
- Glycerol or another rather viscous transparent material with a refractive index close to 1.5.
- A holder for a wedge cell [8, 9] for more accurate measurements.
- A rotatable table for more accurate measurements.

The wedge cell is the cell which has slides in direct contact on one side and a thick spacer consisting of two or three cover glasses on the other. The cell is filled with ordered liquid crystal that forms a thin prism with a very small apex angle $\delta$. The cell is assembled in the same way as cells discussed in the previous subsection, with a spacer on one side only. If the cell is prepared as experimental equipment, it is advised that sides are glued with a sort of inert glue in order to prevent the leaking of liquid crystals. If it is well sealed, the cell could be used for several years.

To obtain properly ordered liquid crystal inside the cell, the surfaces have to be rubbed parallel to the direction of the wedge. Therefore the director in the cell is parallel to the wedge. This geometry guarantees that both refractive indices, for the polarization along the long molecular axes and for the polarization perpendicular to it, remain the same as long the plane of refraction is perpendicular to the wedge. Let us call the cell with this geometry the *parallel wedge cell* (see figure 2.12(*a*)).

*Splitting the light beam with a parallel wedge cell.* The introduction experiment is very simple. A laser beam is shone on the parallel wedge cell as shown in figure 2.13(*a*). Each slide of the cell has parallel surfaces and does not influence the direction of light beams. The refraction of glass does not need to be considered in the analysis of the beams' directions. It is enough to consider the liquid crystal in the wedge cell having the shape of a prism inside the cell. The incident light is perpendicular to one side of a liquid crystal prism. The light does not refract but two beams with different speeds and perpendicular polarizations propagate through the liquid crystal. Both beams refract when leaving a liquid crystal, according to Snell's law. As the incident angle for both beams is equal and the speed of light is different, refraction angles are different and the beams have different directions in the air. The apex angle of the wedge cell is small, but after some distance, a meter or two at least, the light spots indicating two beams are separated enough to be clearly seen on the screen.

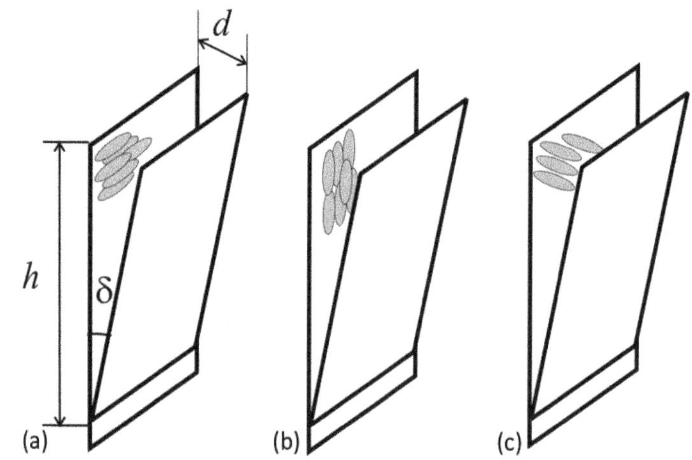


**Figure 2.12.** The three cells needed for detailed studies of refractive indices in liquid crystals. Ellipsoids indicate the type of order in the cell. (*a*) The nematic director is parallel to a glass surface and to the wedge. (*b*) The nematic director is parallel to a glass surface and is perpendicular to the wedge in the parallel wedge cell. (*c*) The nematic in the perpendicular wedge cell director is perpendicular to the glass surface in the homeotropic wedge cell.

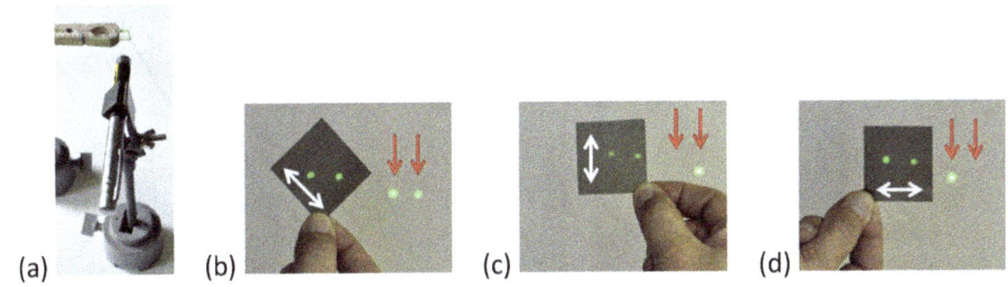


**Figure 2.13.** Determination of the polarization of light after transmission through the wedge cell. Transmission direction is marked by a white double arrow. Arrows indicating light spots and missing spots on the screen serve as guides to the eye. Two light spots on the polarizer are due to reflection and are not important for the experiment. (*a*) The simple setup to observe two dots on a distant screen. (*b*) Closer look at the dots through a polarizer. (*c*) If the transmission direction of the polarizer is parallel to the wedge (i.e. the director) only the light beam polarized parallel to the director and refracted more remains. (*d*) The light beam with polarization perpendicular to the director is refracted less (Photo: Maja Pečar).

The polarization of both beams is verified by an analyser, see figures 2.13(*b*), (*c*), and (*d*). It is clearly seen that polarizations are mutually perpendicular. As a director direction in a cell is known, one easily recognizes that the beam with polarization parallel to the director is more refracted and the other one that is perpendicular to the director is less refracted. The refractive index for polarization parallel to the director is therefore the larger one of the two.

Heating the cell above the clearing point reveals an interesting effect. As the refractive index in the isotropic phase does not depend on the polarization of light, the two light spots on the screen coalesce into one at the clearing temperature. During the cooling the two spots reappear when the nematic phase stabilizes again below the clearing temperature.

*The homeotropic wedge cell.* Repeating the experiment using a homeotropic wedge cell does not yield the same result. As the director is perpendicular to the glass surface, all polarizations have the same speed. This situation is characteristic for propagation (see figure 2.14(*c*)) of light along the optical axis. In nematic liquid crystals, the optical axis is only one; they are uniaxial as the structure is cylindrically symmetric, and the optical axis is always parallel to a nematic director.

However, if the incident light on the prism surface is oblique, the symmetry of the situation is broken. The light having the polarization with a component along the director now has a different, smaller speed, and refracts differently. By increasing the angle of incidence in the plane perpendicular to the wedge, the two spots appear again on the distant screen. For small incident angles the dots are close but become separated at incident angles of 10°. A comparison of two observations—a single light spot for a perpendicular incidence and two spots for an oblique incidence—suggests that the refractive index for a polarization with a component along the director depends on the direction of the light in the material. For a detailed study of the dependence of the refractive index on the direction of light with respect to the optical axis, more effort is required and this is discussed later.

*Detailed measurements of refractive indices.* The ellipsoid of wave fronts in the plane cross-cutting the prism are shown in figure 2.14 for three different wedge cells. In order

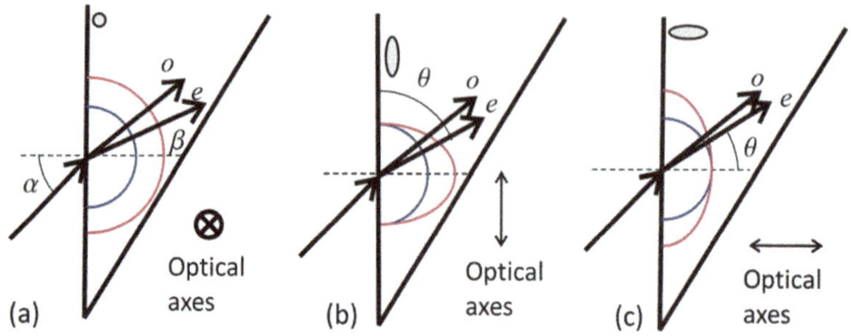


**Figure 2.14.** Wave–vector ellipsoid cross sections relevant for the experiments. Double arrows indicate the orientation of the optical axis. Incident angle $\alpha$ is the same for the three different geometries, while the angle between the refracted beam direction and the optical axis $\theta$ is different. (*a*) In the parallel wedge cell it is always 90° for the geometry of the experiment. (*b*) In the perpendicular wedge cell it is equal to $\frac{\pi}{2} - \beta$, where $\beta$ is the angle of refraction. (*c*) In the homeotropic cell $\theta = \beta$.

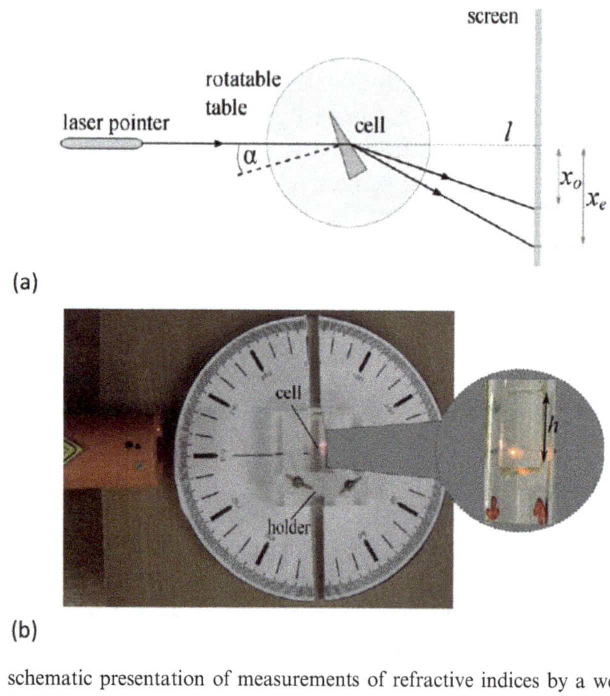


**Figure 2.15.** (*a*) A schematic presentation of measurements of refractive indices by a wedge cell. (*b*) A real setup (Photo: Jerneja Pavlin).

to measure refractive indices dependent on direction in a liquid crystal, the incident angle has to vary in the plane that contains the optical axis. Therefore only the perpendicular wedge cell that is the cell with planar orientation and direction perpendicular to the wedge, (figure 2.12(*b*) and 2.14(*b*)) and the homeotropic wedge cells (figure 2.12(*c*) and 2.14(*c*)) are appropriate for these measurements. Determination of both extreme indices is much simpler. The geometry of a parallel wedge cell, figure 2.12(*a*) and 2.14(*a*), where a director is parallel to a wedge, guarantees that refractive indices do not depend on the incident angle and that the extraordinary refractive index is the largest. Figure 2.15

shows the setup for measuring the distances of the light spots corresponding to the ordinary $x_o$ and the extraordinary beam $x_e$ from the position of the light spot for a direct beam that does not pass a wedge cell. From the data for the distance of the cell to the screen $l$, refraction angles are calculated. As the light beam is orthogonal to the air–liquid crystal interface, it is not refracted. The incident angle on the liquid crystal–air interface is equal to the apex angle of the wedge, $\delta$, which is estimated from data for a spacer $d$ and its distance from the wedge $h$. Usual apex angles $\delta \approx d/h$ of wedge cells are around 1°.

As the apex angle is small and also angles of refraction are small, a calculation of the refractive index is rather straightforward.

$$n_o = \frac{\sin \beta_o}{\sin \delta} \approx \frac{\beta_o}{\delta} \qquad \beta_o \approx \frac{x_o}{l}$$
$$n_e = \frac{\sin \beta_e}{\sin \delta} \approx \frac{\beta_e}{\delta} \qquad \beta_e \approx \frac{x_e}{l}. \tag{2.3}$$

Here $\beta_o$ and $\beta_e$ are refraction angles of the ordinary and the extraordinary light beam. Angles are in radians.

For a detailed measurement of the dependence of the refractive index on a direction in an anisotropic material, wedge cells with planar alignment and a director perpendicular to the wedge (figure 2.12(*b*) and 2.14(*b*)), and homeotropically aligned wedge cells (figure 2.12(*c*) and 2.14(*c*)), are used [9]. The geometry of the experiment is more complicated and the calculation of the refraction index is more tedious, as several refractions have to be considered. But nevertheless, calculation of the refractive index from measured geometrical data is still elementary.

Figure 2.16(*a*) shows the geometry that allows measurements to be taken of refractive indices for angles between the optical axis and the extraordinary beam direction, below 30° and above 60°. The meaning of the symbols is the following: $\delta$ is the apex angle of the liquid crystal prism, $\alpha$ is the incident angle to the incident surface of the prism, $\beta$ is the angle of refraction at the air–liquid crystal interface, $\beta \pm \delta$ is the incident angle at the liquid crystal–air interface and, finally, $\gamma$ is the angle between the incident

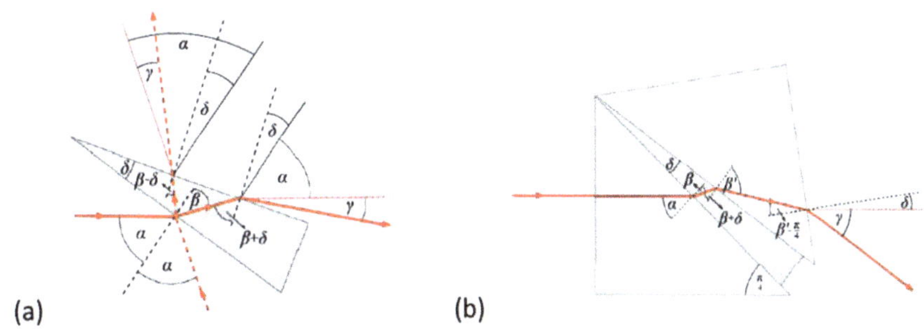


**Figure 2.16.** (*a*) Schematic presentation of the setup for measurements of direction dependent refractive indices. (*b*) Directions close to 45° with respect to the optical axis are achieved by additional prisms and the refractive index matching of the prism and the glass by glycerol. The symbols are explained in the text.

direction and the beam direction after all refractions. The angle $\gamma$ is measured in the same way as previous refraction angles $\beta$, and Snell's law is applied at both surfaces.

$$n = \frac{\sin\alpha}{\sin\beta} \quad \text{and} \quad n = \frac{\sin(\alpha \pm \gamma \pm \delta)}{\sin(\beta \pm \delta)}. \tag{2.4}$$

As $\gamma$ and $\delta$ are small, approximations for small angles are used and the direction of the light beam with respect to the orthogonal to the incident surface $\beta$ is

$$\tan\beta = \frac{\delta}{\delta + \gamma}\tan\alpha. \tag{2.5}$$

The angle $\beta$, which gives the direction of the light beam with respect to the surface normal, corresponds to the angle between the directions of the light and the optical axis $\theta$ in the homeotropic wedge cell. However, $\theta = \pi/2 - \beta$ in the perpendicular wedge cell, see figure 2.14. When $\beta$ is known, the refractive index for the direction of the light beam and its specific polarization inside of the anisotropic material is known, as well as the corresponding light speed.

Equations (2.4) and (2.5) are general equations that are valid for any material, isotropic, uniaxial and biaxial. In uniaxial materials, the dependence of the direction dependent extraordinary refractive index on the angle between the direction of light propagation and the optical axis $\theta$ is

$$n_e^2(\theta) = \frac{n_e^2 n_o^2}{n_o^2 \sin^2\theta + n_e^2 \cos^2\theta}. \tag{2.6}$$

As refractive indices of liquid crystals are large, not all directions $\theta$ inside of the material can be accessed. To overcome this additional prisms are set as shown in figure 2.16(*b*). Two equal prisms (for example, with an apex angle of 90° or 60°) with a refractive index close to the refractive indices of the liquid crystal are placed, as seen in figure 2.17(*a*). The wedge cell is put between the prisms and optical contact is provided by a layer of glycerol between the prism and the wedge glass. The incident light is orthogonal to the prism's side, and as refractive indices for the material of the prism and the liquid crystal do not differ much, the light direction also does not change much. Refraction at the prism glass to liquid crystal interface is therefore much smaller than from the air to liquid crystal. Therefore a value of $\theta$ rather close to 45° is achieved for a perpendicularly incident light to the surface of the glass prism with the apex angle of 45°. However, the apex angle of a prism of 60° provides two additional angles, $\theta$ close to 60° for a homeotropic wedge cell and $\theta$ close to 30° in a planar wedge cell. An example of a measurement taken in our students' laboratory is seen in figure 2.17(*b*) [9].

Three very important phenomena are shown by the set of experiments presented in this subsection. First, the non-polarized beam splits into two mutually orthogonally polarized beams with different speeds and, consequently, different refractive indices. Second, the birefringence, that is the difference between the two refractive indices, is characteristic for any anisotropic material, therefore also for a liquid crystalline state, and it vanishes in the isotropic phase. Finally, at least one of the refractive indices in liquid crystals is direction dependent.

(a)

(b)

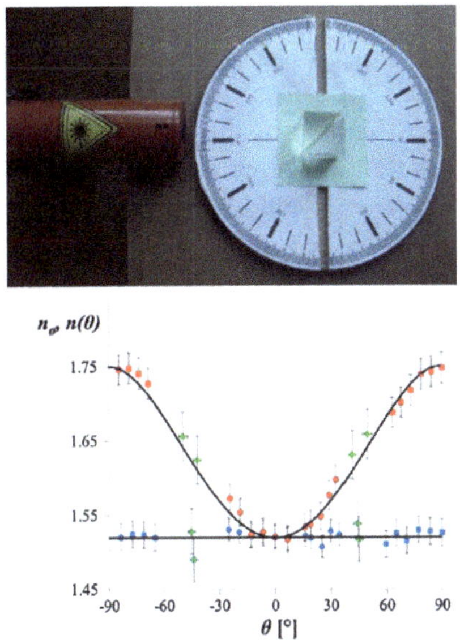


**Figure 2.17.** (*a*) A setup for the measurement of the extraordinary refractive index close to 45° with respect to the optical axis (Photo: Jerneja Pavlin). (*b*) The measured ordinary and extraordinary refractive index for the mixture E18 provided by Sharp. Dots mark the indices measured in the homeotropic cell, squares mark measurements obtained by the perpendicular wedge cell. The two measurements marked by green dots were obtained with the help of external glass prisms. Theoretical expressions for both indices are given as solid lines.

*Tricks of the trade*. The set of experiments starts with the basics, where a double refraction is shown, and continues to elaborate measurements of refractive indices dependent on direction. The experiment at the beginning can be shown as demonstrations.

Putting glycerol between prisms is essential as even a very thin layer of air results in internal refraction within additional prisms. However, the presence of glycerol is a source of additional errors. The glycerol often forms a tiny wedge between the prism and the glass of the wedge cell. In a distance of a few meters, this glycerol wedge could be a cause of a light spot's shift of a few cm. Therefore the accuracy of measurements using prisms by simple methods is rather poor. Still, the trend of the dependence is clearly seen as a measured extraordinary refractive index is found in the gap between the measured values for the two geometries without prisms.

Apart from the first introductory experiment, performing the experiments properly is time consuming. Therefore it is suggested that the whole set of experiments is carried out as project work, where students get different cells and different materials and try to measure the whole dependency of the refractive index, with the derivation of the necessary equations. If the time allocated does not allow for such extensive experimenting, students should measure an extraordinary refractive index for at least a few directions, but the prism sandwich could be left out.

*Suggestions for the lecturer*. The derivation of expressions for the extraction of refractive indices should be left to students as an exercise. Double refraction has no direct application to the understanding of the function of an LCD, it is more a presentation of an optical property. As experiments allow for the relatively accurate measurement of refractive indices, these experiments support the comprehension of optical anisotropy, of values for typical refractive indices of liquid crystals and also of the awareness of the fact that, in any direction, two mutually perpendicular waves exist with different speeds that are direction dependent.

**Additional tasks**

1. Derive an expression for both refractive indices in the planar wedge cell with a director parallel to the wedge. Use this expression for measurements in your experiment.
2. Derive an expression for extraction of the refractive index for an oblique incidence for the same cell as above. Compare the measured values to the theoretical dependence of both refractive indices. Do they change?
3. Derive an expression for both refractive indices in the planar wedge cell with a director perpendicular to the wedge and for the homeotropic wedge cell for oblique incidence. Use this expression to extract refractive indices from your measurements.
4. Make a glycerol optical bridge between the two glass prisms with apex angles of 90°. Measure the shift of the light spot for a normal incidence on the prism to both sides. Does a shift occur? What is a possible reason for a shift? Clean the glycerol from the prism and repeat the experiment again with a new glycerol bridge. Is the shift the same, or different? Why? What happens?
5. Derive an expression for extraction of the refractive indices for the prism–wedge–prism sandwich and calculate refractive indices from the measurements.

### 2.4.3 Elliptical polarization

The comprehension of various states of light polarization is mostly limited to a good comprehension of linear polarization, but comprehension of other types of polarizations, such as circular or elliptical, is often more difficult.

The experiments on the polarization of light given in this subsection have two aims. The experiments can lead to better comprehension of elliptically polarized light and to the understanding of parameters that determine the state. However, an elliptically polarized light is the essence of LCD function. While the setup of LCD guarantees that for the dark state the light passing through the cell retains the polarization, for the bright cell the polarization plane is rotated by 90°. Grey levels are obtained by changing the polarization state of the transmitted light to elliptical. Therefore it is very important that students understand the reasons for elliptically polarized light, how parameters that characterize an elliptical state can be determined and how various properties of liquid crystals can be determined from properties of transmitted light.

**Experimental aims**

- To realize that an anisotropic material between two crossed polarizers allows the transmission of light.

- To determine the polarization state of light passing through a liquid crystal between crossed polarizers.
- To determine the principal directions of light's polarization propagating through the liquid crystal in the cell.
- To study properties of polarized light after transmission through a planar cell.
- To determine the elliptical polarization state and the phase difference of the two orthogonally polarized beams.
- To compare the elliptical states of polarized monochromatic incident light with different wavelengths.

**What is required for the experiments**

- A planar liquid crystal cell.
- Two or three lasers with different wavelengths (red, green, blue).
- Two polarizers.
- A detector for the intensity of light, a luxmeter or a photoresistor.
- An optical bench and rotatable polarizers.

As long as we are working with nematic liquid crystals, the most evident results are obtained with the planar cell with the thinnest spacers, or even without them. The test that the cell is appropriate for the experiments is that colours are seen between crossed polarizers and are more or less homogeneous (figure 2.18).

The initial experiments are of an observational nature. The cell is put between crossed polarizers with a rubbing direction oriented at a general angle with respect to the transmission direction of the polarizers. A grey area, or colours at the place where a liquid crystal is, can be seen; the proof that light is transmitted although the polarizers are crossed.

Next, the cell is rotated between the crossed polarizers until the cell seems dark. The transmission directions of the polarizer $P$ and the analyser $A$ are marked on the cell. These two directions coincide with the polarizations of both beams that the incident light splits into. The coordinates system axes $x$ and $y$, that will be used later for the analysis of the measured signal, are assigned to these two directions.

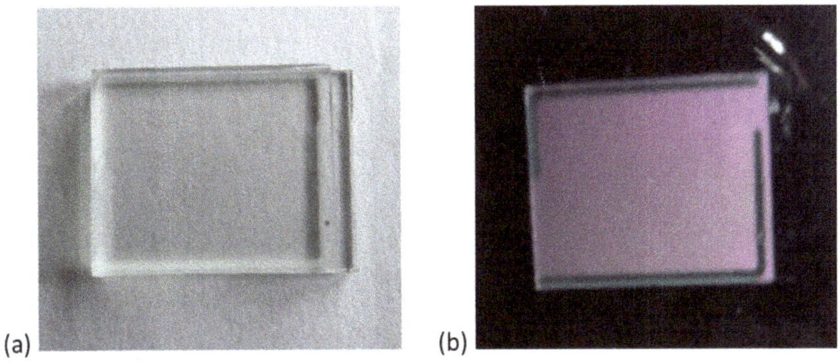

**Figure 2.18.** (*a*) Professional cell provided by INSPEC filled with a liquid crystal (RTN by Merck). (*b*) The same cell between crossed polarizers. Homogeneous colour indicates very good and homogeneous order.

These experiments can be done by putting the equipment on simple stands, but an optical bench makes experiments more controllable and reliable. Therefore we suggest using an experimental set up on a bench, as seen in figure 2.19. The setup is the same as for more detailed measurements of Malus' law, the only difference being that the cell with an ordered liquid crystal is inserted between polarizers.

On the optical bench, the directions of a coordinate system are determined more accurately by shining the laser on the position in the cell that is marked as well ordered. The marking of an ordered area is done in the following way: the cell is placed between two crossed polarizers and it is rotated between them. For two mutually perpendicular orientations the cell is seen as mostly dark. The largest dark regions are marked and the laser beam in the experiment should pass through one of them.

An experiment that yields the same results, the determination of polarization directions, can also be done under the microscope. For two specific orientations the cell will be mostly very dark. For these orientations a direction of long molecular axes correspond to the transmission direction of one of the polarizers. Dark areas the size of the laser beam cross section can be marked on the surface of the cell using this method as well. These areas are used in more accurate measurements later.

Next, the cell is rotated with a rubbing direction of under 45° with respect to the transmission directions of the polarizer and the analyser. The monochromatic light is shone through the first polarizer, the liquid crystals and the second polarizer that serves as an analyser. Lasers or laser pointers have a well defined wavelength and serve well as monochromatic light sources. The analyser is rotated by 5° or 10° in one step and the transmitted intensity of light is measured (figure 2.20). The measured intensity of the transmitted light is also presented graphically and various suggestions for graphical presentations are considered figure 2.21.

By performing the experiment, students learn to recognize properties of elliptically polarized light. They find the intensity of the transmitted light varies depending on the angle between the analyser and the polarizer direction. If a different wavelength laser beam is used, the elliptically polarized transmitted light has different properties. How can such a behaviour be explained and how can such measurements be used in the determination of the properties of the liquid crystal ordered in the cell?

(a) (b)

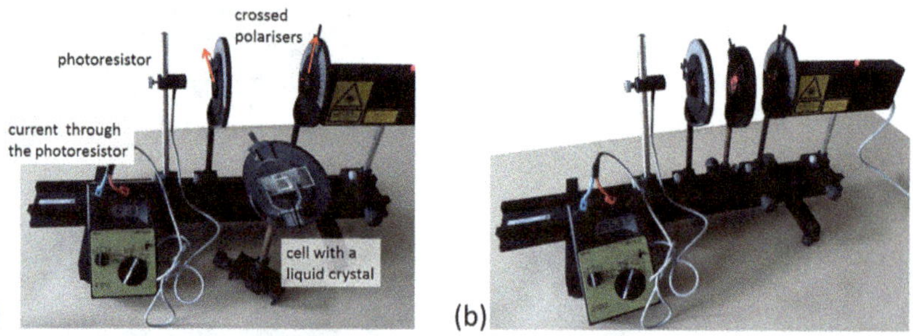


**Figure 2.19.** (*a*) The laser, rotatable polarizer and analyser, and photoresistor for the measurement of intensity of transmitted light. The ammeter shows that the current through the photoresistor is very low if polarizers are crossed. (*b*) The planar cell is inserted between crossed polarizers. The ammeter shows a significantly higher current indicating a much higher intensity of light.

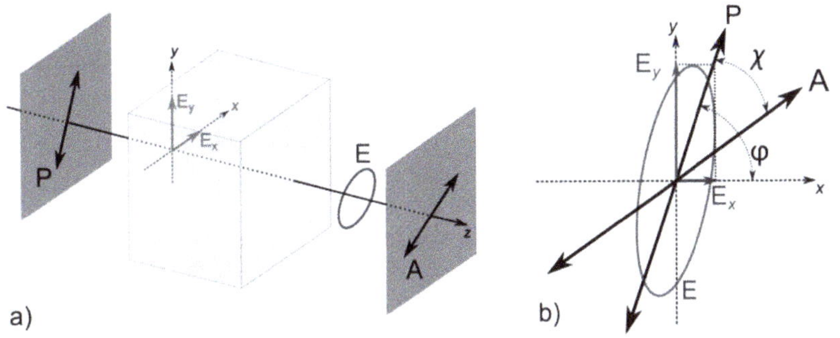


**Figure 2.20.** (*a*) Coordinate system with respect to the sample. (*b*) The geometry for the analysis of the transmitted light.

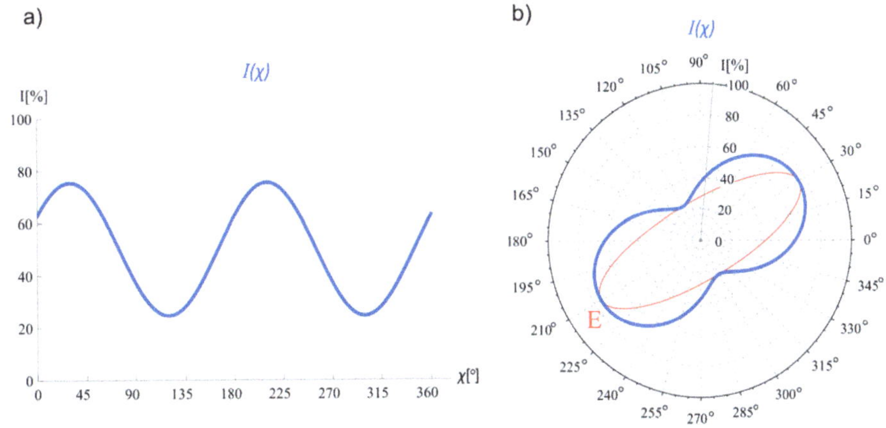


**Figure 2.21.** Graphical presentations of intensity of transmitted light depend on the angle between transmission direction of the polarizer and the analyser $\chi$. (*a*) Usual Cartesian graph. (*b*) Polar graph. The red ellipse inserted into the polar graph is the presentation of the elliptical polarization state. It presents the curve drawn by the tip of the electric field vector at a fixed position during one period of elliptically polarized light wave.

These are actually the basics of optical studies of liquid crystals and the basis for applications in various devices.

The perpendicularly incident light splits into two beams that have the same direction in a material but a different speed. Here we consider the direction of a wave vector and not the direction of an energy flow, characterized by a Poynting vector. The reason for this sloppiness is simple. The wave vector defines refraction of light beams and the phase differences between two partial beams. In the experiments discussed here, the cells filled with liquid crystals are used and they are always so thin that a space separation between the two beams does not occur. Therefore the phase shifted beams interfere after transmission and result in an elliptically polarized wave. Planar alignment of the sample guarantees that principal axes of the dielectric tensor of the liquid crystal in the cell correspond to the rubbing direction and to the direction perpendicular to it. It makes the analysis and the comparison of theoretical results and the measurements rather simple. The geometry for the analysis is given in figure 2.20(*b*). The electric field in the perpendicularly incident linearly polarized light propagating along the $z$ axis in an anisotropic material is given by

$$\vec{E}(z, t) = \begin{Bmatrix} E_x\cos\left(\omega t - k_x z\right) \\ E_y\cos\left(\omega t - k_y z\right) \end{Bmatrix}$$

$$= \begin{Bmatrix} E_0\cos\varphi\,\cos\left(\omega t - k_x z\right) \\ E_0\sin\varphi\,\cos\left(\omega t - k_y z\right) \end{Bmatrix} \tag{2.7}$$

where $k_x$ and $k_y$ correspond to wave vectors of partial beams polarized along the $x$ and the $y$ directions that correspond to eigenvectors of the dielectric tensor. The phase of beams after transmission through an anisotropic material of thickness $d$ is $k_x d$ and $k_y d$, respectively. The phase difference between the two beams is

$$\delta = k_x d - k_y d = \frac{2\pi}{\lambda_0}(n_e - n_o)d \tag{2.8}$$

where $\delta$ is the phase difference and $\lambda_0$ is the wavelength of light in vacuum. The refractive index $n_e$ corresponds to the speed of propagation for the polarization parallel to the director and $n_o$ corresponds to the speed of propagation for polarization perpendicular to it. The choice of coordinate systems makes this correspondence straightforward.

Finally, the intensity of transmitted light is given as an average intensity that corresponds to the projection of the electric field to the transmission direction of the analyser. The dependence of the intensity of transmitted light with respect to the rotation of the analyser given by $\chi$ is

$$I = \frac{1}{2}\varepsilon E_0^2\left(\cos^2\chi - \sin 2\varphi \sin 2(\varphi - \chi)\sin^2\frac{\delta}{2}\right). \tag{2.9}$$

For proper extraction of the properties of the studied liquid crystal, the measured data is fitted by the expression given in equation (2.9) for three different wavelengths at least. All wavelengths should be fitted with the same product $\Delta nd$, where $\Delta n = n_o - n_e$. The task is not so straightforward in reality, as the phase difference yields the same result for $\delta$ and $\delta \pm k2\pi$. Therefore students have to find the range of phase difference for the studied transmission. The colour of the sample is very helpful and will be considered in the next subsection.

*Tricks of the trade*. Alternatively, cells can be ordered from various providers. Such cells are professional and guarantee very good ordering. For studies of polarization states it is advised to have a set of professional cells. They last a long time and the problems with good alignment that are found using hand-made cells are avoided. However, it is also of benefit to graduate students entering the field to gain experience with ordering problems. Several misinterpretations of results can occur if the samples studied are not well ordered.

*Suggestions for the lecturer* Experiments that lead to the determination of optical properties of anisotropic materials are rather demanding and are appropriate for advanced optical or soft matter labs. Our experiences were that younger students often experiment according to cookbook suggestions and do not very often develop a conceptual comprehension. Not to mention the fact that the tasks here should follow previous thorough lectures on optics of anisotropic materials. More about various experiments to determine the elliptical state and properties of anisotropic materials with foils and tapes instead of liquid crystals is presented in the paper [5].

**Additional tasks**

1. Set up the system on the optical bench. Measure the intensity of the transmitted signal for three different orientations of the polarizer: parallel to the director in the cell, under 45° with respect to the director, and for any other angle with respect to the director. Draw Cartesian and polar graphs from the data.
2. Compare points for the same measurements on both graphs. For example, the intensity of light measured, and the angle between the transmission directions of the polarizer and the analyser $\chi = 30°$, 45° and 60°.
3. Fit the polar graph with varying $\delta$. Extract the birefringence $\Delta n$ for a known thickness of the cell. Repeat the measurement and the procedure for a laser with a different wavelength of light. Does the birefringence extracted from the measurement depend on the wavelength? Is it possible? It may be necessary to reinterpret your measured data.

### 2.4.4 Colours

Observations of a liquid crystal between crossed polarizers and a liquid crystal between polarizers oriented with general transmission directions very often reveal colours. Colours usually vary from place to place in the cell and a homogeneous colour is a mark of good alignment. Why do the colours appear? The answer is hidden in the equation for the phase difference. The intensity of transmitted light depends on the elliptical state of the polarization and the orientation of the long axis of the ellipse with respect to the transmission direction of the analyser. In figure 2.22, the effects of two parameters, the orientation of the incident beam polarization given by $\varphi$ and the phase difference between the two partial beams given by $\delta$, are clearly seen.

While the first parameter $\varphi$ is easily changed by the orientation of the polarizer defining the incident polarization, several properties contribute to the second parameter $\delta$. Let us repeat the equation for the phase difference and let us write equation (2.8) in a more general form.

$$\delta = \frac{2\pi}{\lambda_0}(n' - n'')d. \tag{2.10}$$

Here, three influences are clearly recognized: the wavelength of the light $\lambda_0$, the thickness of the sample $d$ and the difference between the refractive indices for the two perpendicular polarizations of the two beams $(n' - n'')$. The expression written in this form is valid for any direction of light and for any anisotropic material, uniaxial and biaxial. Unfortunately, neither refractive index is very easy to determine for general directions of incident light [5]. The easiest to influence is the wavelength of the incident light, $\lambda_0$. This is the reason why lasers or laser pointers of different wavelengths were used for experiments in the previous subsection.

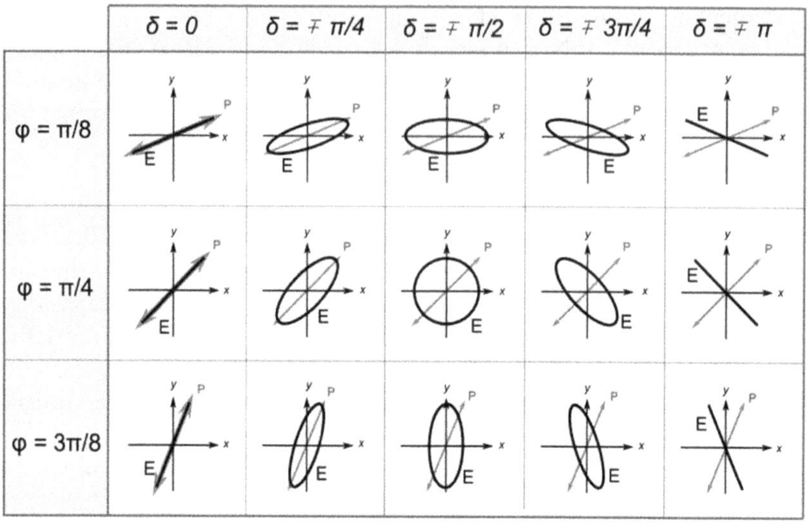


**Figure 2.22.** Effects of the phase difference $\delta$ and the incident plane of polarization $\varphi$ on the orientation and the shape of a characteristic ellipse. Note that by changing $\delta$ the long axis of the characteristic ellipse rotates, except for the polarization direction forming 45° to the principal polarization directions in the anisotropic material.

However, if the sample is illuminated by white light, all wavelengths corresponding to the visible light spectrum are present simultaneously. As they have different wavelengths, they have different elliptical states after transmission and are absorbed at different rates by an analyser. The spectrum of the transmitted light is different to the spectrum of the incident light which is detected as a colour by the naked eye.

The colour phenomena associated with the birefringence for a known sample thickness is important information on the studied samples. Several points can be extracted from Michel–Levy charts, see figure 2.23 [10].

**Experimental aims**

- To find conditions in which planar cells filled with liquid crystals reveal colours.
- To find conditions in which homeotropic cells filled with liquid crystals reveal colours.
- To analyse how colours change with the cell thickness.
- To discuss reasons for different colours in the same cell.
- To make a simple analysis of the defects.

**What is required for the experiments**

- Planar cells of various thicknesses, starting with cells without spacers.
- Homeotropic cells of various thicknesses, starting with cells without spacers.
- Two polarizers with known transmission directions.
- A wedge planar cell with a thin spacer at one side (a layer or two of food wrapping foil).
- A microscope.
- The Michel–Levy birefringence chart.

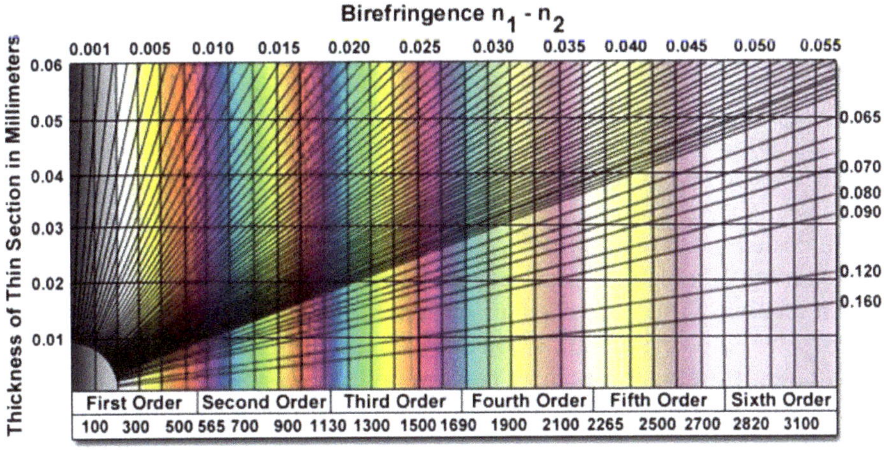


**Figure 2.23.** The Michel–Levy birefringence chart for colours of samples between crossed polarizers. The horizontal axis presents retardation, defined as $(n' - n'')\, d$. The vertical axis on the left gives the thickness of the sample. Lines originating in the left lower corner connect colours that have the same birefringence $(n' - n'')$. By comparing the colour of the sample to the colour on the Michel–Levy chart, the birefringence is determined providing the order is known. Note that for larger birefringences, scales have to be renormalized.

Students observe cells of various thicknesses between crossed polarizers. For planar alignment, a very thin cell already reveals the colour. Unlike professional cells, the thickness of thin planar hand-made cells is not usually known precisely. They also reveal several colours. The reason for this is the very high birefringence of liquid crystals, as the difference $\Delta n$ is as much as 0.3 for planar alignment.

The thin planar oriented cell is observed under a microscope. The colours usually seen are in the left part of the Michel–Levy chart. Yellow and blue colours indicate the observation of the first order colours. Applying a little pressure on the cell, colours change due to the change of thickness. Thicker cells reveal only magenta and green colours; their presence usually indicates higher orders in the Michel–Levy chart. Colours also change if the cell between the crossed polarizers is tilted.

Homeotropic cells are different. Between the crossed polarizers the cell is dark as the light is propagating along the optical axis and there is no retardation between the beams. Using thicker cells, it is easy to realize that the refractive index in nematic liquid crystals depends on the direction of light propagation. If the homeotropic cell between crossed polarizers is slightly tilted, it becomes transparent or even coloured.

For more demanding laboratory studies, a thin planar wedge cell is prepared. The whole Michel–Levy set of colours appear between crossed polarizers when observing the wedge cell due to the continuous changes in cell thickness. If the thickness of the spacer at one side is known, the local thickness in the cell is easily calculated and as orders (multiples of $\pi$ phase difference) are clear, the birefringence is straightforward to read.

*Suggestions for the lecturer.* If the motivation for the introduction of liquid crystals is understanding LCDs, it is strongly advised that the colours of liquid crystals in cells between crossed polarizers are not considered before the function of

LCDs is discussed. Colours of pixels are obtained by coloured filters and the colour of the liquid crystal between crossed polarizers have no relation to the colours on an LCD screen whatsoever. Experiments presented here make a strong impact on students and very often students simply believe that colours they observed in experiments presented in this subsection are the reason for colours in the screen. Although it would be possible to construct the screen based on phenomena discussed in this subsection, the phenomena used in regular screens is more sophisticated and will be discussed in chapter 3. Special structural changes of order within the screen provides a more reliable and easier control of the colours.

In addition, we suggest that phenomena associated with the colours of anisotropic materials between polarizers are studied separately using simple anisotropic materials like tapes, transparencies etc. They are more accessible and easier to manipulate. Experiments for studying these phenomena are discussed in subsection 4.4.1.

**Additional tasks**

1. In some cells that are not well ordered, a colour changes from one position in the cell to another. Rotate the cell between crossed polarizers and try to suggest possible reasons for the changes of colour.
2. Calculate the birefringence for different places in the cell assuming the thickness is constant. What are the possible reasons for the variation of colour if one assumes a constant magnitude of the order and a constant cell thickness?
3. Observe the wedge cell between crossed polarizers in red and blue light. Use red and blue colour filters. Connect the colour of the wedge cell at specific places where dark fringes appear for a specific colour of incident light. Consider the subtractive colour scheme as the cell absorbs some wavelengths more than the others, as discussed in subsection 3.2.

## Bibliography


[1] Liberko C A and Shearer J 2000 Preparation of a surface-oriented liquid crystal an experiment for the undergraduate organic chemistry laboratory *J. Chem. Educ.* **77**(9) 1204–5

[2] Patch G and Hope G A 1985 Preparation and properties of cholesteric liquid crystals *J. Chem. Educ.* **62**(5) 454–5

[3] Sluckin T J, Dunmur D A and Stegemeyer H 2004 *Crystals That Flow: Classic Papers from the History of Liquid Crystals* Liquid Crystals Book Series (London: Taylor & Francis)

[4] Repnik R, Mathelitsch L, Svetec M and Kralj S 2003 Physics of defects in nematic liquid *Eur. J. Phys.* **24** 481–92

[5] Pečar M and Čepič M 2015 Conoscopic figure: a complex consequence of a not so simple phenomenon *Eur. J. Phys.* **36** 010514

[6] Hecht E 1998 *Optics* 4th edn (Reading, MA: Addison-Wesley)

[7] Born M and Wolf E 1999 *Principles of Optics* 7th edn (Cambridge: Cambridge University Press)

[8] Shenoy K D 1994 Measurement of liquid crystal refractive indices *Am. J. Phys.* **62** 858–9

[9] Paulin J, Vaupotič N and Čepič M 2013 Direction dependence of the extraordinary refraction index in uniaxial nematic liquid crystals *Eur. J. Phys.* **34** 331–44

[10] http://www.olympusmicro.com/primer/techniques/polarized/michel.html. Accessed: 2014–09-09

# Chapter 3

# The liquid crystal display

The most common application based on liquid crystals is a liquid crystal display (LCD). Students and everybody else see LCDs every day in computer screens and mobile telecommunication equipment, such as smart phones and tablets. A students' familiarity with these applications makes the idea of understanding how the LCD works an attractive motivation for studying liquid crystals and can be used as a light motive for the topic.

How does an LCD operate? All LCDs are constructed similarly. Students can get an initial idea of how an LCD operates by observing a magnified screen. The basic unit, the pixel, is recognized and the colour math rules are discussed. Colour math is a topic that can also be studied independently in optics, related to colours and vision. The structure of the screen allows the brightness of each part of the pixel in each pixel to be independently addressed. This is achieved by changing the structure of liquid crystal by an application of external electric field. The light becomes polarized when it enters the screen, changes its state of polarization depending on the structure of the liquid crystal in the screen, and is finally partially or completely absorbed before leaving the screen structure. Different polarization states of light are absorbed differently and this is how the brightness of the pixel's parts is manipulated. The colour of the light is more simply provided, by colour filters. The structure of the screen is seen in figure 3.1. Constituent elements of the screen that have been discussed, or will be discussed later, are marked.

The amount of data that has to be processed in a short time for satisfactory quality of moving images is tremendous. Several additional requirements have to be met in order to guarantee a clear picture, viewing angle etc, however, they will not be discussed further as the advanced technology related to the solutions of these problems goes far beyond the purposes of this book.

## 3.1 Screen colour math

To familiarize students with the structure of a screen they should observe a magnified one. An approximate idea can come from an observation through an ordinary

doi:10.1088/978-1-6270-5300-6ch3  

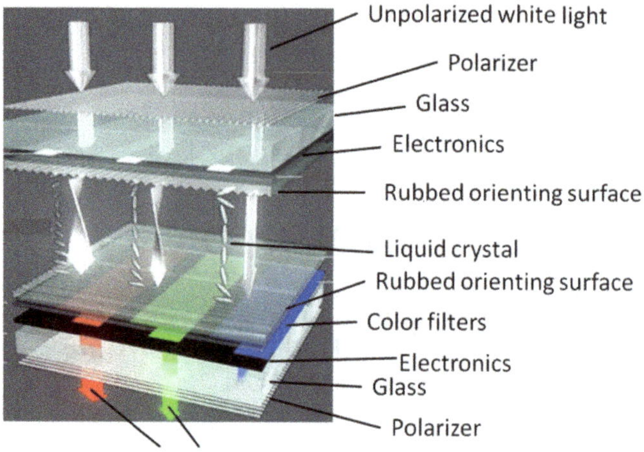


**Figure 3.1.** A simplified structure of one pixel, adapted by [1].

magnifying glass. Although details of the structure cannot be seen clearly, the student will see how it is composed and that the screen is not uniform as surfaces of everyday objects are. For a more detailed observation, stronger magnification is needed. For elementary observations students can put small drops of water on the surface of the screen and then observe the screen through the water droplet using a magnifying glass or a watchmaker's loupe. Greater magnification is achieved by a USB microscope [2]. The details on the screen are clearly seen and allow the pixel to be recognized as a basic unit that defines colours. The colour math is introduced by simple observations, and the numbers relating to the colours on the screen are given meaning.

The screen is composed of pixels in which the intensity of red, green and blue emitted light is manipulated independently. The viewing angle for observation of the pixels is much smaller than the resolution of an observer's eye and the three elementary colours, red, green and blue, mix. The colour math for additive colour mixing has the following rules, see table 3.1.

To obtain any colour, the intensity of light transmitted from the back side of the screen through a colour filter is controlled. When one defines a colour of a background or a colour of an object, the colour scheme presented in figure 3.2 appears as one of possibilities. The colour is defined by dragging a cross through the coloured area and simultaneously the colour coordinates given below the colours change. The colour can also be defined in the opposite way; numbers between 0 and 255 are entered for each primary colour: red, green and blue. What do these numbers mean? The 0 means that the light is completely absorbed and no light is transmitted through the corresponding colour filter. On the other hand, 255 means that the pixel and its internal structure of liquid crystals is such that the light absorption is as small as possible. So for a single pure primary colour, only red, green or blue has a value for determining transmission, 255, and the values for the other two colours are zero. Similarly, for elementary mixtures that form the subtractive elementary colours: yellow, magenta and cyan, transmission factors for two colours, as presented by the

**Table 3.1.** Rules for additive mixing of coloured lights. The magnified screen for the corresponding colours is given in pictures.

| | | | | | | |
|---|---|---|---|---|---|---|
| red | + | green | | | = | yellow |
| red | + | | | blue | = | magenta |
| | | green | + | blue | = | cyan |
| red | + | green | + | blue | = | white |

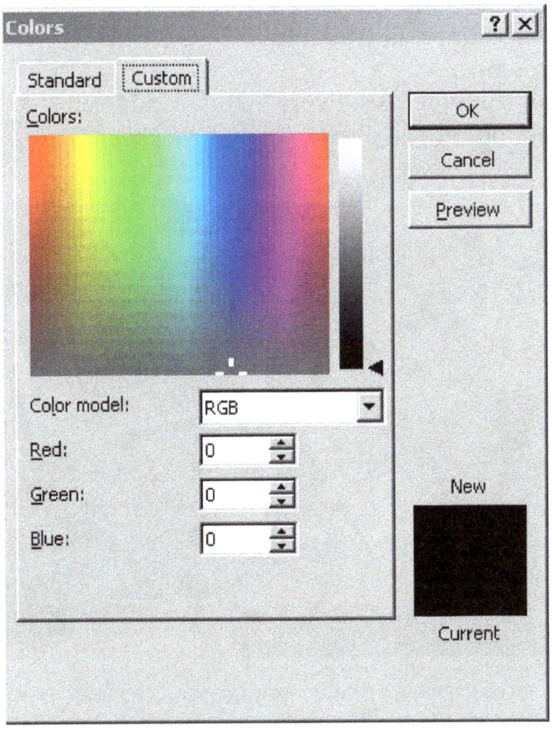


**Figure 3.2.** Determination of a colour on the screen.

colour math in table 3.1, are set to 255 and the third one is 0. Finally, for white, all the light is transmitted and all factors are set to 255. All other colours are defined by varying the transmission factors for each primary colour. For example, orange is obtained as a mixture of green and red that has lower intensity than for yellow and even lower for brown [3].

The aim of the activities presented in this section is for students to familiarize themselves with how colour on the screen is formed. This activity could serve as an introduction to liquid crystals, and provide good preparation for the later study: what is the role of liquid crystals in controlling the transmission of light through the screen?

**Experimental aims**

- To realize that the screen is not uniform but composed from constituent parts that have three different colours.
- To realize that the colours are produced by three elementary coloured light sources with controlled emission of light.
- To learn colour math.
- To define and recognize the pixel as a fundamental part of an LCD screen.
- To become familiar with the composition of various colours on computer or TV screens.

**What is required for the experiments**

- A portable computer.
- A magnifying glass. Glasses should have different magnifications, if available.
- A watchmaker's loupe, if available.
- A USB microscope.
- An additional computer (see figure 3.3), if available.
- Various LCDs on mobile phones, digital watches, etc.

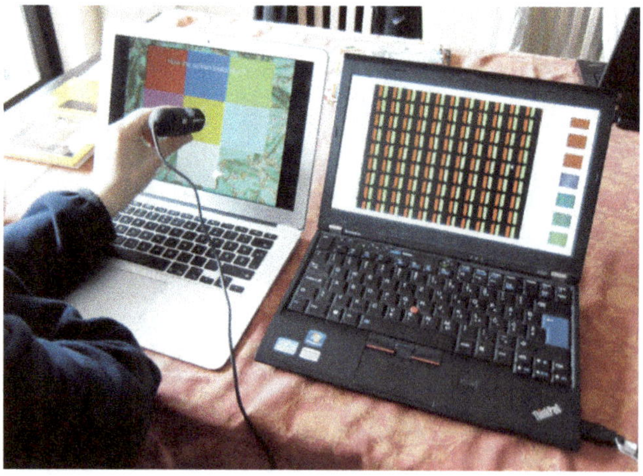

**Figure 3.3.** Observation of colours using a USB microscope. Use two portable computers, if possible; one screen to be studied and one screen on which to observe the magnification.

**Figure 3.4.** Example of a screen for the observation of additive mixing of colours. Pictures are added to 'pure' colours, allowing students to follow the changes in the emission of light defining colours by changing the position of the microscope.

Prepare a picture that has areas of elementary colours (red, green and blue), elementary mixtures (yellow, magenta and cyan), white, and a part with a general photo that has brighter and darker areas of various colours. An example is seen in figure 3.4. The picture should also be sent by email to students so that it can be studied on mobile phone screens too.

An LCD from a portable computer is observed through a magnifying glass. A mobile phone screen should be observed in the same way. The glass is positioned several cm away from the screen, where the magnification is the greatest. The observed screen should be in the focal length distance. Greater magnification is obtained using a watchmaker's loupe but students will need to get used to the loupe.

Next, the USB microscope is used. The use of a microscope should be explained, then students are allowed to discover the rules of colour math by observing the largest possible magnification of the screen allowed by the USB microscope. They should use the preprepared image containing pure and mixed colours and observe it using the USB microscope at various places on the screen.

Finally, students should open the colour fill option in a program such as PowerPoint or similar and click through 'Background fill' → 'More colors' → 'Custom' → 'RGB'. They should study the numbers associated with the pure and mixed colours. Students should also purposely change the numbers that give transmission factors for pixel elements and observe the corresponding colours on the magnified screen.

*Tricks of the trade.* Magnification of the screen using a USB microscope is much easier if two computers are used. The program for presenting enlarged images runs on one and the observation runs on another. It is possible to have the magnification and the observed part on the same screen but it is more difficult and the magnification is not so clearly observed, particularly if the screen is small.

*Suggestions for the lecturer.* The colours on the screen are composed of three elementary colours. That much can be observed by simple magnifications, but to see more detail, greater magnification by a USB microscope is needed.

A USB microscope is first introduced as a demonstration of the USB microscope as a camera. By adjusting the focus the lecturer proceeds to closer and closer objects, finally reaching the magnification of objects at a predefined distance when the microscope protection is in close contact with a surface on which a studied object is lying. Two available magnifications are shown. The lecturer also demonstrates the effect of rotation of the microscope on the image on the screen. Next, the lecturer explains the manipulation of the USB microscope; how to change the focus, i.e. the magnification, and how to change the illumination. The lecturer should also discuss the circumstances for which additional illumination is required. Next, students are allowed to play with a microscope to familiarize themselves with its functions. Students usually study available objects such as skin, hair, various clothing fabrics, pencil drawings, and printed pictures. One could also prepare some interesting objects for observation. We suggest pieces of plant tissues, a piece of a sponge and other materials that have interesting structures on a microscopic level.

When students are familiar with the USB microscope they are asked to observe a yellow paper, for example, and then to observe the yellow colour on the screen. Not everybody has come across the difference before. After this introduction, students proceed to study the construction of the colour math and to study how colours can be changed.

**Additional tasks**

1. Measure dimensions of pixels on your portable computer screen or on your mobile phone. Use the magnification of a ruler to estimate the magnification of the USB microscope.
2. Familiarize yourself with colours on the screen. Construct a Powerpoint slide showing a number of different objects which are uniformly coloured. Exchange the slides between groups and find the RGB that define the colours on the screen. Also try the opposite: choose a colour on the screen and try to guess its RGB components.
3. Try to recognize different parts of the picture on the screen by observing the magnification only. Work in pairs.

## 3.2 Printer colour math

This subsection is not about liquid crystals but it is a nice complement to activities about display colours. Printers are everyday devices too, however, the pictures they print for us at home, in textbooks and in journals, are taken for granted. Students should be asked if there is a difference between the colour on screen (red, green, blue components) and the colour in print. Printers have magenta, cyan and yellow inks accompanied by a black one. The following activities offer a natural extension to what was learned in subsection 3.1.

**Experimental aims**

- To study the magnification of printed colours, compare them to the colours on the screen, and deduce the printer colour algebra.
- To study magnifications of printed pictures and deduce how brighter and darker colours are achieved.
- To study magnifications of pictures printed in colour in magazines.

**What is required for the experiments**

- A portable computer.
- An additional computer, if available.
- A USB microscope.
- A printed picture containing pure elementary additive colours (red, green, blue) and elementary subtractive colours (magenta, cyan, yellow) similar to the one in figure 3.4.
- A printed picture of squares with the same colour in different variants (darker and brighter, with additional components of other colours such as adding a small amount of blue to the red, and similar). The same picture should also be available as a Powerpoint slide or similar at the screen.
- Printed colour pictures from magazines, textbooks with colour pictures and similar, drawings using crayons and colour pencils.

Print the picture similar to the one presented in figure 3.4. The same picture should be available to students on the computer screen. Students observe the same place on the picture on the screen and on the print version with a USB microscope, and compare the structures. The colours red, green and blue consist of a single bright colour on the screen but are made up from two colours on the print. Red consists of magenta and yellow dots, green consists of cyan and yellow dots and blue consists of magenta and cyan dots. However, magenta, cyan and yellow are printed with dots of the specific colour only. Darker colours are obtained by the addition of black dots and brighter colours are achieved by leaving parts of the white paper between the dots uncovered. While the three colours in a pixel make the whole plethora of colours by varying the brightness of pixel parts, the structure of printed colours is based on different areas being covered by colours, with the addition of black or white. Why is that so?

The printed colours are subtractive. The picture in the eye is formed from the reflected light. Usually pictures are observed in white light and the observed colour is obtained by subtracting colour components from the white light in reflection. The magenta pigment, for example, reflects wavelengths around blue and red. The cyan pigment reflects wavelengths around green and blue light. A small piece of cyan pigment reflects in any direction the light with green and blue components, the rest is absorbed. The reflected light falls on a piece of magenta pigment, the remaining green components are absorbed but blue is still reflected. If the sequence is reversed, the blue components are reflected as well. The intensity of blue components is relatively larger in the reflected light than it is in the incident light and the part of the picture is seen as blue. The colour math associated with the pictures observed in

**Table 3.2.** Rules for subtractive mixing of reflected coloured lights. Examples of colours are magnified prints. Note: the black should be obtained by printing magenta, cyan and yellow at the same place, but using black ink is much cheaper.

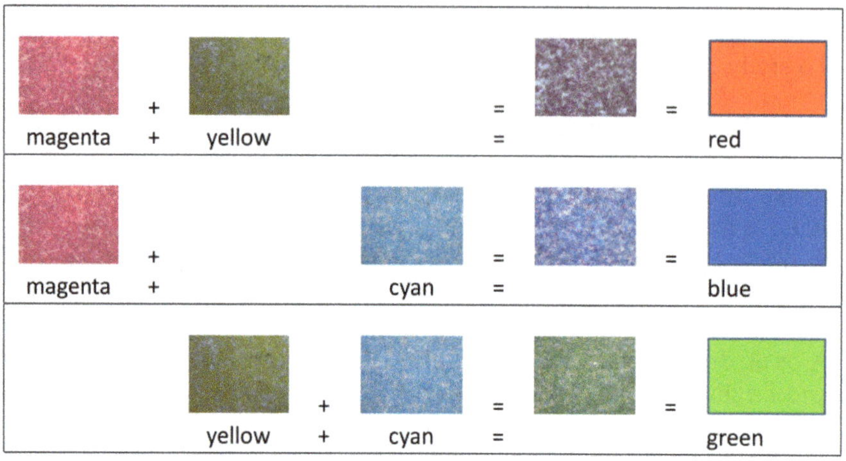

| | | | | | | | | |
|---|---|---|---|---|---|---|---|---|
| magenta | + | yellow | | | = | | | red |
| magenta | + | | | cyan | = | | | blue |
| | | yellow | + | cyan | = | | | green |

reflection is called the subtractive colour math and is exactly the opposite to the additive colour math presented in subsection 3.1. To illustrate this, the subtractive colour math is presented in table 3.2.

For economical reasons, printers do not obtain black by mixing elementary printing colours. However, older printers had only three colours and some of us still remember that black from a colour printer was not really black.

*Tricks of the trade*. Examples for observation should be printed in low resolution. Good modern printers have too high a resolution sometimes and, under the microscope, the structure of the colour is not seen.

For observation of printed examples, the illumination in the USB microscope should be used.

**Additional tasks**

1. Study various printed pictures under different magnifications. Work in pairs. Try to find the magnified part on the picture.
2. Study prints of a single colour with various hues, brightnesses and saturations. How are those properties manifested in magnification?

## 3.3 Electric field

In the previous subsection students have learned that the image on the screen consists of pixels with three elementary colours (red, green and blue). The brightness of the pixel elements further determines the colour actually seen on the screen. The whole construction of the part containing liquid crystals operates as a shutter that dims the transmitted light. The transmitted light passes the colour filter providing the colour of the corresponding part of the pixel [4].

**Experimental aims**

- To prepare liquid crystal cells with conductive surfaces that allow voltage to be applied across the cell.
- To study transmission of light in dependence of the applied voltage.
- To construct a model pixel.

**What is required for the experiments**

- Glass coated by a conductive layer.
- Aluminium foils as contacts.
- Pieces of transparency for spacers.
- A stand to support model cells.
- Two polarizers.
- A source of variable voltage.

The availability of glass covered by a conductive layer is crucial for this activity. Such glasses are usually available in research laboratories where researchers make their own cells with the desired properties for research purposes.

The glass covered by the conductive layer is cut into pieces the size of typical object glasses. When assembling the cell, the conductive layers should be in the inner side of the cell. The glass surfaces on the coated side are rubbed in the desired direction to allow for planar alignment. For demonstration purposes, it is enough that the rubbing direction on both surfaces is parallel, but students may also play with different orientations. A better model of a pixel is made if the surfaces are rubbed almost perpendicularly, that is, the rubbing directions should form an angle of approximately 80°.

Careful measures have to be taken so that the surfaces do not come into direct contact when the voltage is applied (and also otherwise) because the cell would not be operative. This is prevented in the following way: the spacers in the form of a thin transparency should be large enough to cover a relatively wide area between the treated glasses. The upper glass is shifted with respect to the lower glass and the aluminium foil is placed between the transparency and the upper glass on one side and the transparency and the lower glass on the opposite side. The transparency prevents contact with the opposite glass. The foils serve as conductors and allow the voltage to be applied. The whole cell is heated with a hair dryer and the liquid crystal is introduced by capillary action. It is also possible for glasses with spacer and foil contacts to be prepared separately. Again, both glasses are heated, a drop of liquid crystal is placed on one glass and is covered by the second one.

An example of such a cell can be seen in figure 3.5.

The voltage is applied. When the voltage is high enough, elongated molecules in the middle of the cell start to rotate with the long axis parallel to the external electric field.

*Tricks of the trade*. It is very important that there is no contact between the cell surfaces. In addition, the cell should have thin spacers as one would like to achieve rather high electric fields with low voltages, provided by simple means such as batteries. It is better to use AC voltage. The phenomenon is quadratic and does not distinguish between the parallel or antiparallel directions of the electric field. As ions are always present in liquid crystals, they slowly move towards electrodes in a DC

**Figure 3.5.** The planar cell with contacts that allow the external electric field to be applied. The cell is filled only partially, the liquid crystal is present at bright places only. The rubbing direction forms an angle of 45° with the transmission directions of the polarizers (Photo: Maja Pečar).

field and screen the electric field. The movement is suppressed in a low AC voltage and prevents the screening effects.

*Suggestions for the lecturer*. The structure of the screen and the function of pixels should be considered before colours are observed between polarizers. Students often mix up the two phenomena, the colour of the liquid crystals between polarizers and their function as a simple control of a more or less transparent pixel window.

**Additional tasks**

1. Analyse the transmission of light theoretically. Assume that a typical pixel in the screen is a few μm thick and the liquid crystalline molecules in the screen form a structure as discussed at the beginning of this chapter.
2. Measure the intensity of the transmitted light depending on the applied voltage for a planar cell. Calculate the electric field where the intensity of light starts to change. Note that the polarizer and the analyser should be under 45° for the best observation.
3. Assemble a twisted cell and measure the intensity of the transmitted light depending on the electric field. How does the intensity of the signal differ? Note that the the polarizer and the analyser above and below the cell should be parallel to the corresponding rubbing direction.

## Bibliography

[1] Picture adapted from http://www.merck.de/de/unternehmen/entdecke_merck/lcd_explorer.html. Accessed: 2014-09-09

[2] Several providers offer relatively cheap USB microscope. For example: http://www.adafruit.com/products/636. Accessed: 2014-09-09

[3] Erjavec M and Vaupotič N 2006 Bottle model of colour vision with the colour brown as an example *Eur. J. Phys.* **27**(3) 611–20

[4] Pečar M, Pavlin J, Susman K, Ziherl S and Čepič M 2011 Hands-on experiments for demonstration of liquid crystals *Proc. 8th Int. Conf. on Hands-on Science Focus on Multimedia* **2** 38–42

# Chapter 4

## Accompanying concepts

This chapter considers phenomena important in the understanding of liquid crystals, without using liquid crystals. The main purpose of additional experiments without liquid crystals is to observe the visual evidence of the phenomena that they provide. When studying liquid crystals, most of the experiments already assume that concepts such as order, anisotropy, polarization of light etc are already understood and can be used for the deduction of properties and internal structure of liquid crystals. Some students are good abstract thinkers and establishing a link between an observation of a real phenomenon and the theoretical expression describing the same phenomenon presents no substantial problem. However, for some students visual assistance, not in the form of cartoons or pictures, is very helpful. Our interviews with the best students and renowned researchers have shown that simple hands-on experiments illustrating methods they use every day in the laboratory are very welcome and have prompted them to see their research studies from a different perspective.

Three major concepts are considered: the concept of anisotropy and the order parameter, visualized and measured through very simple mechanical models, the presentation of optical properties using wood and microwaves, and the optics of easily accessible everyday materials.

Most of the experiments can be used in advanced optic courses as easy illustrations of phenomena, but experiments illustrating an order parameter have meaning within a course on liquid crystals.

### 4.1 Anisotropy as a concept

Anisotropic properties are properties that depend on the direction of external stimuli. Such properties are often come across in everyday life although students are seldom aware of them. Let us quote few examples.

Running through a vineyard is easy if one runs parallel to the lines where vines are growing (figure 4.1(*a*)). If one wants to cross a vineyard perpendicularly to the vine lines, running is impossible. One has to bend over every second meter to get

doi:10.1088/978-1-6270-5300-6ch4

**Figure 4.1.** Everyday examples of structures that are not the same in all directions. (*a*) Vineyards have vines 'ordered' in one direction. (*b*) Wood has fibres 'ordered' in one direction. (*c*) The knitted pattern has two well defined directions with different properties: the knitting direction and the line direction. All those structures give rise to different properties in different directions.

through the wires etc. So, the speed of running depends on the direction of running. The vineyard is anisotropic as the properties related to transport depend on the direction of transport through the vineyard.

Another familiar example is wood (figure 4.1(*b*)) [1, 2]. Sawing wood in different directions with respect to wood fibres is different. The nicest cuts are obtained perpendicularly to the wood fibres, but parallel to the fibres the wood is inclined to cleaving. The absorption of water also depends on the direction of wood fibres [3]. Wood easily absorbs water along its fibres but absorption of water perpendicularly to them is not so efficient. We say that wood is anisotropic for absorption or that there is an anisotropy in absorption in wood.

Finally, an everyday example that is suggested for a more detailed study later, is knitted fabrics (figure 4.1(*c*)) [4]. Everybody knows that a T-shirt or a pullover extends in one direction more than in the other, and that it shrinks in one direction differently to in the other during washing and drying. There are also several other properties that are associated with knitted fabrics and their patterns that mean the response of the fabrics to external stimuli is different in different directions, which is typical for anisotropic materials.

What do the examples quoted above have in common? Each of the examples has a structure that is different in one direction to the other. The vineyard has long lines of vines (figure 4.1(*a*)), the wood has parallel fibres (figure 4.1(*b*)), the knitted fabric has lines in one direction and the orientation of purls is perpendicular to the line (figure 4.1(*c*)). Such structures that are not the same in all directions are called anisotropic, in contrast to isotropic materials in which no variation of structure with direction exists. Correspondingly, as the structure is different in different directions, it is consistent that the materials with anisotropic properties react differently to stimuli in different directions.

The anisotropic structure is clearly seen in the examples discussed above. In research it is usually the opposite. In experimental results, one recognizes anisotropy in responses and from such indications the possible structures are suggested. From an educational point of view, it is useful to provide students with experience of examples where the anisotropy is visually evident. Such experiences allow the easier deduction of anisotropic structure from experimental results later. Two easy and straightforward experiments provide such experience. The first experiment shows the essence of anisotropic properties—the extension of a knitted pattern is different if the force is applied perpendicularly than parallel to the knitting direction. The ratio between extensions depends on the material and on the pattern itself. The second experiment shows another important phenomenon. The response of an anisotropic material is not parallel to the direction of the external stimulus, in general. If the force is applied to a knitted fabric in a general direction, the fabric's extension is in a different direction to that of the applied force.

### 4.1.1 Principal extension directions of knitted fabric

Let us first discuss why a knitted pattern can be used as a model for demonstrating anisotropy and its properties. A knitting pattern has two distinct directions. One is the knitting direction, in which the knitted structure grows during the knitting. The other is the direction perpendicular to the knitting direction that corresponds to the lines of the knitted pattern (figure 4.1(*c*)). It is also well known that the number of lines forming a square can be very different from the number of stitches forming the line. For example, the three samples with different patterns in figure 4.2, differ significantly in the number of stitches per cm and in the number of lines per cm (table 4.1). Although the two numbers, the number of lines and the number of stitches, differ and are a hint that the properties may be anisotropic, they cannot be used as a measure for anisotropy. The numbers indicate that the properties might be different in different directions only. The number of lines and the number of stitches per cm cannot be influenced by external stimuli. Therefore we have to search for another property which can be measured and depends on the direction of the applied external stimulus, for example, the applied force.

Everybody wearing sweaters knows that the knitted material stretches in the knitting direction differently to the direction perpendicular to it, along the lines. So, the elastic coefficient should depend on the direction of the applied force. Let us define a quantity that can be measured and evidently shows anisotropy in the properties of a piece of knitting. A comparison of the different patterns presented in figure 4.2 made of the same wool or other threads is presented in table 4.1. It is clearly seen that the number of lines per cm and the number of stitches per cm vary

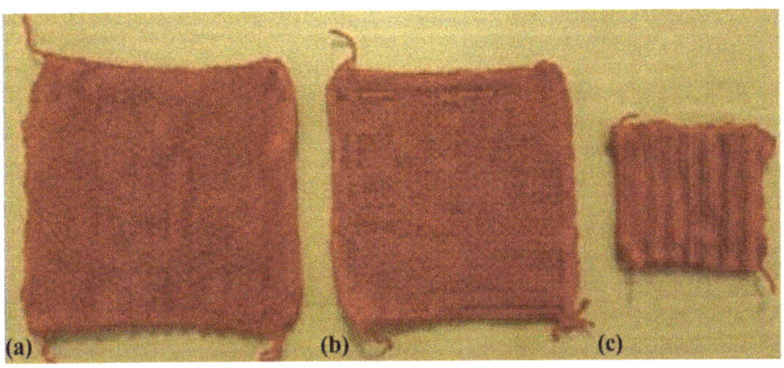


**Figure 4.2.** Examples of knitted patterns with 30 stitches. (*a*) Rice: 1 knit 1 purl in the first line, 1 purl 1 knit in the second line, and so on. (*b*) All purls in both directions of knitting. (*c*) 2 knits, 2 purls, a well known pattern for the ends of sleeves.

**Table 4.1.** Properties of the three samples given in figure 4.1(*b*).

| | Rice | Purls | 2 purls, 2 knits |
|---|---|---|---|
| Lines ($cm^{-1}$) | 4.4 | 5.5 | 3.5 |
| Stitches ($cm^{-1}$) | 2.5 | 2.7 | 3.8 |
| Width $d$ (cm) | 12.3 | 11.0 | 6.8 |
| Length $l$ (cm) | 12.0 | 10.9 | 7.0 |
| Elastic coefficient along the knitting direction $k_{\parallel}$ (N $cm^{-1}$) | ≈0.5 | ≈0.7 | ≈2.0 |
| Elastic coefficient along the line $k_{\perp}$ (N $cm^{-1}$) | 0.6 | 1.5 | 0.5 |

significantly from pattern to pattern and so do their appearance. The rice pattern has a rather homogeneous structure, while the other two patterns have structures evidently different in one direction to the other.

One property of the samples that is easy to measure is an elastic coefficient in Hook's law approximation:

$$k = \frac{\Delta F}{\Delta l}, \tag{4.1}$$

where $k$ is the elastic coefficient as in Hook's law, $\Delta F$ is an increase in the applied force and $\Delta l$ an increase in the sample length. As samples of different patterns usually differ in size, the measured elastic coefficients have to be normalized to allow for comparison of elastic properties of different patterns.

$$\kappa = k\frac{l}{d}. \tag{4.2}$$

Here $\kappa$ is the pattern elasticity normalized to the length $l$ and the width $d$. Measured data for samples in figure 4.2 are found in table 4.1 and serve as an orientation to the lecturer.

**Experimental aims**

- To show that a knitted fabric extends in the knitting direction differently to how it extends in the line direction.
- To measure elastic constants of knitted fabrics for two mutually perpendicular directions, along the knitting direction and along the lines for different patterns.
- To calculate and to compare the elasticity of different patterns knitted from the same thread.

**What is required for the experiments**

- Various knitted patterns prepared using the same number of stitches and the number of lines so that the sample is approximately square.
- Light hard thin plastic sticks.
- Weights.
- Ruler and stands to hang the samples up.

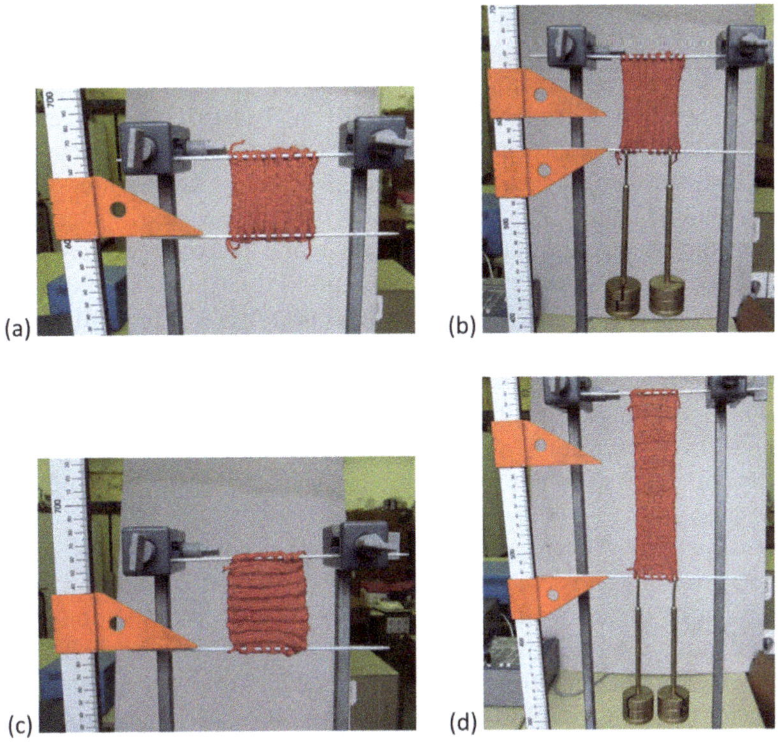

**Figure 4.3.** The knitted pattern called 'knit two purl two' used for the open ends of knitted pullovers (around the neck, at the end of the sleeves and at the bottom). (*a*) Relaxed sample prepared for loading in the knitting direction. (*b*) The same sample extended along the knitting direction. (*c*) Relaxed sample prepared for loading along the lines. (*d*) The same sample extended along the lines. Note: one extension is significantly larger than the other under the same weight.

Linear dimensions (width $d$ and length $l$) of relaxed samples are measured. The samples are prepared for measurements in the following way: plastic sticks are introduced into the cloth close to the edges, along the lines of the sample as in figure 4.3. One stick serves to support the sample and the other on the opposite side of the sample serves as a hanger for weights. The sample is fixed between two stands using the stick and the weights are loaded on the lower stick. Weights of the same mass have to be attached along the stick in such positions that the stick is horizontal. Then the extension is measured and the elastic coefficient and the pattern elasticity is calculated from the measured data. The procedure is repeated for different samples with different patterns. After the elastic coefficient for the force applied along the knitting direction is measured, the sticks are removed and introduced again along the knitting direction. The extension for the same force as in the first experiment is measured and from the measured data elastic coefficients and the elasticity is calculated.

The experiment is very simple and straightforward and clearly shows that the response of the material, and consequently the properties of the material, can depend on the direction of an external stimulus, the force in the studied case. The suggested experiment is performed in a distinguished geometry, the force is applied along the knitting direction and perpendicular to it. Almost all common knitted patterns are such that in these two directions the extension is parallel to the applied force. Directions in which the response of the material has the same direction as the external stimulus are called principal directions and for such directions properties also have distinguished values that are called principal values. For example, in anisotropic materials three directions exist in which the electric displacement vector has the same direction as the applied electric field and dielectric constants that relate the electric displacement vector, and the electric field in these directions are called principal dielectric constants. The dielectric constant is a tensor. It is similar in studies of elastic properties of knitted patterns. Extending the pattern perpendicularly to the knitting direction and to the lines at the same time has no meaning in those patterns, although it is theoretically possible. The knitted patterns can be considered as two dimensional and only two directions exist in which the extension has the same direction as an applied force. The elastic properties of a knitted pattern are also a tensor.

*Tricks of the trade.* There are several tricks that complicate this easy experiment. First, nowadays it is not easy to find an experienced knitter. If a novice tries to knit a sample, thread would not be homogeneously strained and the sample itself would not be homogeneous. Old pullovers or T-shirts can be used as a source of samples. Samples are cut from the back or from upper parts of sleeves, where the cloth is not worn out too much. In this way several groups can compare the same pattern. However, comparing different patterns is more complicated as a pullover knitted from the same thread but with different patterns in different parts has to be found. Samples from different pieces usually lack control of variables as the properties of a thread (the thickness and the material), and the stress applied by a knitter or by a knitting machine during the knitting, significantly influences the properties. Still, the comparison can be done but one has to bear in mind that properties can only be assigned with certainty to the specific sample with a certain pattern made of a certain thread. For the pattern itself, only a general conclusion could be drawn.

The extension of the sample cannot be described by Hook's law. When weights are added the range of forces in which the extension is not proportional to the applied force is easily reached. Usually the material comes to 'saturation' and the extension is almost constant for larger forces. If patterns are prepared from elastic threads, the problem of non-linearity in extension is not so crucial. Nevertheless, if the lecturer wants to show the differences in extension for different directions of forces only as a demonstration, this discussion is not relevant. The lecturer should find proper weights so that the phenomenon is evidently shown and that is all. For more detailed measurements, students should undertake a proper study that includes the measurement of extension for several applied forces and draw the corresponding diagram, similar to the one presented in figure 4.4. The graph may serve in a debate about the definition of the elastic coefficient. If the average coefficient for smaller forces, when the extension still linearly depends on the load, is taken, results are usually acceptable for a laboratory exercise.

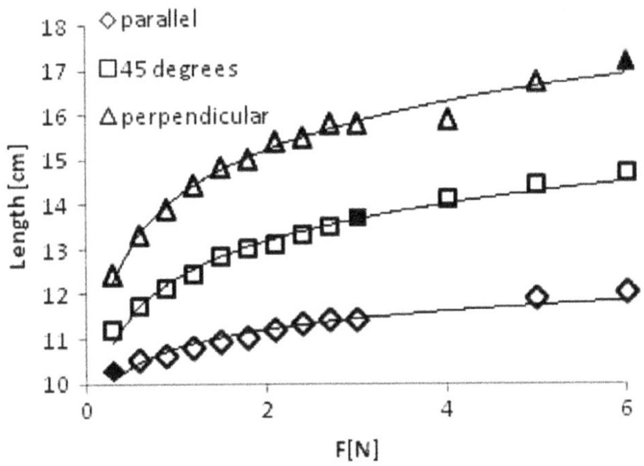


**Figure 4.4.** Measurement of the extension in dependence on force for different directions of applied force. It is clearly seen that the dependence is initially linear but that it soon stops increasing.

*Suggestions for the lecturer*. Experiments can be done as a demonstration or as a laboratory. If experiments are used for demonstration purposes, the patterns studied should be prepared in duplicate. The two samples with the same pattern should be loaded with the same weight and shown at the same time to allow visual comparison of the response of the material in the two principal directions.

**Additional tasks**

1. Measure principal elasticities for equal patterns made from threads with different properties. In which way they are similar, and in which way do they differ?
2. Measure principal elasticities for some additional patterns. For which kind of clothes were the studied patterns used? What were the properties of those clothes? Why were such patterns chosen?

### 4.1.2 Extension under force with a general direction

One of characteristics for phenomena related to anisotropy is the response to an external stimulus having a vectorial character that is not in the same direction as the stimulus itself. A few examples are: heat flows are not parallel to the temperature gradient in anisotropic materials in general, induced polarization is not parallel to the external electric field, and there are several others. The same is true for knitted patterns. The extension has a different direction to that of the applied force, in general. Not only the direction, but also the magnitude, of the response depends on the direction of the external stimulus. In the subsection 4.1.1, only properties in two specific directions were studied. Here we extend our study to forces applied in a general direction with respect to a knitted direction.

**Experimental aims**

- To show that a knitted fabric extends differently if the force is applied in different directions.

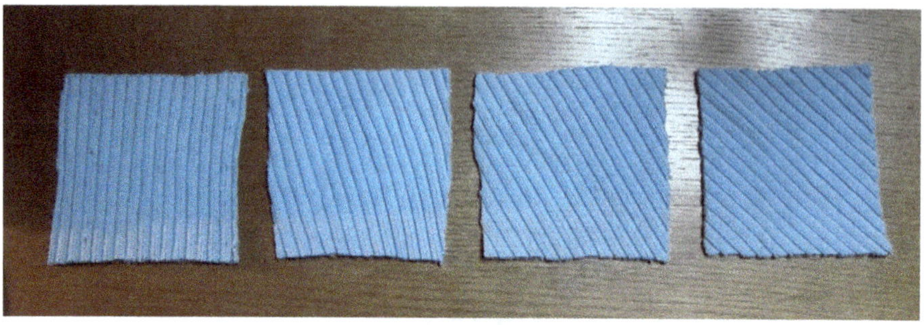

**Figure 4.5.** The samples were cut from the back of an old cardigan. Angles of cutting with respect to the knitting direction are from left to right; 0°, 15°, 30° and 45°. Note that the rectangular shape of the fabrics (except the one cut at 45°) allows for two measurements. For example, the sample cut at 30° with respect to the knitting direction is also used for the measurement of applied force at 60° with respect to the knitting direction.

- To measure the dependence of the elastic constant of a knitted fabric on the direction of the applied force.
- To calculate the elasticity of the same pattern in different directions.
- To observe that the knitted fabric extends in a different direction to the applied force.
- To measure the angle between the direction of the extension and the applied force.

**What is required for the experiments**

- Knitted patterns of a rectangular shape of the same size, cut from the same knitted cloth in various directions (as in figure 4.5).
- Light hard thin plastic sticks.
- Weights.
- Ruler and stands to hang the samples up.

The linear dimensions (width $d$ and length $l$) of relaxed samples are measured. The samples are prepared for measurements in the same way as in previous experiments. Weights are loaded and the extension is measured. The elastic coefficient and the pattern elasticity is calculated from the measured data. The procedure is repeated for different samples that have been cut in different directions. For each direction of applied force, the angle between the free edge of the sample and the applied force direction (the vertical) is also measured.

This set of experiments is very simple and straightforward and clearly shows the response of materials, and consequently the properties of materials, depend on the direction of the external stimulus, the force in the studied case.

The experiment nicely models another phenomenon typical for anisotropic materials. The response of a material to an external stimulus has a different direction to that of the stimulus itself, in general (see figure 4.6), if the property linking the stimulus and the response is a tensor.

*Tricks of the trade*. The angle of the applied force is measured with respect to the knitting direction or with respect to the line direction. It depends on the choice, and

**Figure 4.6.** The extension is parallel to the applied force, if the force is (*a*) parallel to the knitting direction or (*b*) perpendicular to it. (*c*) For other directions of applied force the extension has a different direction to that of the force.

both choices are equivalent. Therefore each sample can be used for the measurements related to two different directions of applied force, only the sample has to be rotated by 90°. However, if the original piece of fabric is large enough, it is advisable that samples with the same knitting directions are cut twice and measurements for the two directions are done with each sample in one direction only. The knitting fabric is inclined to memory effects that can influence the results of the measurements.

*Suggestions for the lecturer*. For demonstration purposes, all loaded samples should be displayed at the same time. This will allow students to experience the effect visually and compare different directions of applied force directly.

As a laboratory exercise, a thorough measurement of an extension with respect to the magnitude of the applied force is advisable, as the extension does not linearly depend on the force. Repeating several measurements of the same type manually is boring and demotivating. Therefore each group of students may measure only one or two samples (samples cut in two different directions) and the dependence of the direction dependent elasticity will be reproduced from the measurements contributed from many groups that have, cumulatively, studied all the angles of applied force.

**Additional tasks**

1. Calculate the dependence of the elasticity on the direction of the applied force. Note that the resulting extension is due to extensions in both principal directions.
2. Compare the calculation of the direction dependent elasticity to the direction dependent extraordinary refractive index in uniaxial anisotropic materials, for example, liquid crystals.
3. Measure and compare direction dependent elasticities of different patterns.

## 4.2 Toothpick model of an order parameter

Let us start this section with a brief summary of the order parameter philosophy.

An order parameter is a measure of order in liquid crystals. For different phases, different order parameters are defined. The most simple order parameter is a nematic order parameter in its simple scalar form. More information is provided by a tensor order parameter [5], but for an introduction to the fundamental concept called the order parameter, a simple scalar nematic order parameter $S$ is satisfactory.

Macroscopically, a phase transition from nematic to isotropic phase under heating is recognized visually as a clearing of an opaque liquid into a clear transparent liquid. A transition occurs at a well defined temperature called the clearing temperature. Properties of opaque liquids are significantly different to the properties of clear transparent liquids. The opaque liquid phase, when it is restricted to a space which does not exceed 200 to 400 μm in one dimension at least, reveals optical properties typical of anisotropic crystals.

Microscopically, the reason for anisotropy in optical properties is in the microscopic ordering of elongated molecules. A schematic diagram presenting a nematic phase and an isotropic phase is shown in figure 1.3. Ellipsoids are a representation of elongated molecules, which rotate quickly around their long molecular axes. The main difference between the liquid crystalline nematic phase and the isotropic phase is an average orientation of long 'ellipsoid' axes or, in reality, long molecular axes. As the order is the main distinction between the isotropic and the nematic phase, the order needs a measure. How should such a measure be introduced?

It should satisfy the following requirements:

- The order parameter has to reflect the symmetry of the system. The measure should not depend on the rotation of a sample around any axis. The measure should reflect the up–down symmetry of the nematic phase.
- The order parameter should be 0 in an isotropic phase.
- The order parameter should be 1 if molecules are ideally ordered, that is if all long molecular axes are parallel.

The order parameter $S$ for a nematic phase is introduced as an average of the second Legendre polynomial $\langle P_2(\Theta)\rangle$, calculated for every single molecule used, to the angle $\Theta_i$ measured with respect to the average orientation of long molecular axes—the director. Why is $\langle P_2(\Theta)\rangle$ chosen?

If the order of long molecular axes is important, the measure of the order parameter should be associated with an angle formed by the molecule with respect to the average direction of long molecular axes, as a cluster of molecules may orient in any direction. A naive idea would be to propose an average $\langle \cos \Theta_i\rangle$ calculated over the sample. As for any direction in the space chosen, one would find as many molecules oriented along that direction as molecules oriented in exactly the opposite direction. We say that up–down symmetry on the microscopic level exists. To include this symmetry in the prospective order parameter, $\langle \cos^2 \Theta_i\rangle$ is a better choice. But the average of $\langle \cos^2 \Theta_i\rangle$ in a disordered phase, in which all directions are equivalent and equally occupied, is equal to 1/3. To obtain the value of order parameter 0, a constant equal to 1/3 is subtracted. And finally, as, for an ideally ordered system, the value of $\langle \cos^2 \Theta_i\rangle = 1$, the difference $\langle \cos^2 \Theta_i\rangle - 1/3$ has to be renormalized. The order parameter is therefore defined as

$$S = \frac{3}{2}\left(\left\langle \cos^2 \Theta_i \right\rangle - 1/3\right) \tag{4.3}$$

calculated as an average over all molecules.

The order changes with the temperature and the order parameter, as a measure of this order, changes as well. The lower the temperature, the larger the component of the internal energy in the free energy determined by various intermolecular interactions, and the weaker the component related to the molecular mobility, to rotations and translations. As a consequence, attractive van der Waals interactions result in the alignment of long molecular axes and the order parameter increases upon lowering the temperature.

Although researchers use the concept of order parameter in debates almost subconsciously, its meaning is not so straightforward. One often forgets phenomena that underlie a single number given for $S$, like a distribution of molecular directions that indicates several molecules can be oriented significantly out of an average direction. This clearly becomes evident if one checks cartoons of liquid crystals from various webpages, where the order of elongated molecules is almost always highly exaggerated. As students obtain information through illustrations as well as the theoretical definition, the concept is often only vaguely understood.

Experiments presented later should fill this gap and allow students to get some feeling for the meaning of the values.

### 4.2.1 Determination of order parameter

An illustration of the order parameter using toothpicks was first proposed in [6]. The experiment uses toothpicks as models for elongated molecules in a nematic liquid crystal. The model is actually two dimensional and the order parameter has to be defined according to the averages valid in two dimensions, $\langle \cos^2 \Theta_i \rangle = 1/2$. The order parameter in two dimensions is therefore

$$S_{2D} = 2\langle \cos^2 \Theta_i \rangle - 1 \tag{4.4}$$

and this equation is used for the calculation of the order parameters in the toothpicks model.

**Experimental aims**

- To measure the order parameter for a set of toothpicks.
- To analyse how the number of toothpicks involved in the experiment influences the experimental results.
- To study distributions of toothpicks with respect to the average direction of the set.

**What is required for the experiments**

- 20–40 toothpicks.
- Squared paper and pencil.
- A goniometer.
- Excel or excel-like program.

The experiment should give students an idea about the meaning of the nematic order parameter and distributions that are behind it. The order parameter is a

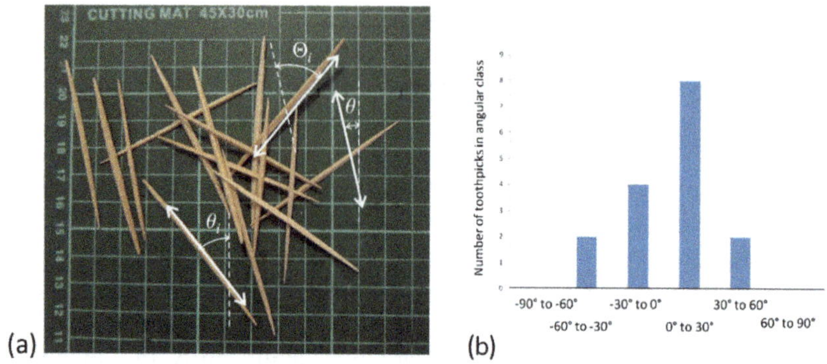


**Figure 4.7.** (*a*) Set of toothpicks after they have fallen. Definitions of particular angles $\theta_i$, the average direction $\theta$ and the particular angle with respect to director $\Theta_i$. (*b*) Distribution around the director for an example with a high order parameter $S_2 = 0.72$.

'statistical' property, that is, its value is well defined only if the number of molecules is very large.

Toothpicks model elongated molecules in a liquid crystal. The direction cannot be associated with a toothpick, because toothpicks are symmetrical. Molecules forming a liquid crystal are not symmetrical to the same extent, but there are as many molecules oriented in one direction as in the opposite direction, on average. The effects of molecular symmetry average out and the properties are the same as if the molecules were symmetric. The symmetric toothpicks are a good approximation of nematic molecules in this aspect.

The experimental procedure is the as follows. A bundle of 20 to 40 toothpicks is lifted 10 to 15 cm above the table. The toothpicks should be parallel to the table surface. Drop the bundle on the paper. The activity is much simpler if toothpicks fall onto squared paper. The result is seen in figure 4.7(*a*). What can be noticed immediately? The toothpicks are more or less oriented along the same direction. In the first step, the angle formed between the toothpick and one chosen direction on the paper is measured. The sense of the angle is included in the sign. Here, squared paper is helpful. The measured angles, with respect to an arbitrary chosen direction (with respect to one of the sets of parallel lines on the squared paper, for example), $\theta_i$ for each toothpick, are inserted into the table and the average angle $\theta$ is calculated. A direction given by the average angle is a model of another concept used in liquid crystals—the orientation of the average long molecular axes, called a *director*. Also, for the set of toothpicks on one piece of paper, one could say that their director indicates the average direction of the toothpicks.

In the next step, from each measured angle the director angle is subtracted and the value $\Theta_i$ for each toothpick is obtained. The angles are now measured with respect to the director and these angles are entered into the calculation of order parameter $S_2$. The average of the expression $2(\cos^2 \theta_i - 1/2)$ is calculated and the value obtained gives the two dimensional order parameter for the set of toothpicks.

In the next step, a distribution of angles within a bundle is considered. The range from $-90°$ to $+90°$ is divided into classes with a width of about $+20°$ around the director. Angles $\Theta_i$ are distributed into those classes and the distribution is presented as a bar graph (as in figure 4.7(*b*)). The measured values of $\Theta_i$ for the bundles dropped from the same height, measured by all groups in the classroom, are merged together and the order parameter of this larger sample is calculated and compared to the values for the order parameters from separate groups.

The order parameter is an abstract concept and, considering the structures of liquid crystals, it is often not clear what the distribution actually is that is the origin for the order parameter. Very often schematic diagrams present tremendously exaggerated order in presentations of liquid crystalline phases, which later results in an erroneous impression of the order. Furthermore, students and often also researchers unconsciously draw conclusions on the basis of the simplified impressions from such diagrams. Therefore the activity here allows students to realize that, even for large values of order parameters, many molecules can be found that are oriented out of group order and that most of the molecules are far from parallel to each other.

The distribution of the toothpicks' orientations in the larger set is presented by a bar graph in figure 4.7(*b*). It can clearly be seen that the distribution for 20 toothpicks is not really informative. While in distributions of smaller sets only general ideas of behaviour, like 'more toothpicks have orientations closer to the director then perpendicular to it', the distribution for larger number of toothpicks shows the shape too. It is also evident that different distributions may yield a similar number of order parameters. From this activity, students should realize that large numbers in statistics are important. Therefore it is suggested that the results from several groups are combined into one distribution. Students should not forget that a relevant point about toothpicks is the angle $\Theta_i$ of a specific measurement with respect to a director. The control variable for all groups should be the height of the fall. All groups have to apply the same height in order for the combined results to be meaningful.

It is also easy to make students aware of another point. The average direction of the toothpicks is influenced by the initial direction of the bundle. So different groups get different orientations for a director, but in the calculation of the order parameter this orientation has no effect. The order parameter reflects that the space is isotropic and elongated molecules of a liquid crystal may orient in any direction. However, the initial direction in which the bundle was held acts like an external stimulus that gives the bundle a preferable direction. The procedure of orientation of the liquid crystals is rather similar. Weak external forces, like the treatment of the surface, determine the direction average for long molecular axes and the consequent director in the sample.

*Tricks of the trade.* There are several types of toothpicks found on the market. Bamboo toothpicks have a perfect cylindrical shape but they roll easily and the angles are often difficult to measure as slight movements of the paper or small unconscious touches of the toothpicks may change their position and orientation. For this experiment, non-cylindrical flat or hand-made toothpicks of general shapes are better.

*Suggestions for the lecturer.* The toothpick experiment can be carried out by one student, in pairs, or even groups of three are acceptable. Larger groups do not allow every member to be actively involved.

The experiment can be done as a hands-on homework. However, it is advisable for the analysis of the results to be done using the combined results of several people/groups, during organized exercises.

### 4.2.2 Drop height dependence of the order parameter as a model of the phase transition

This experiment is simple but demands a little more time and effort. Students model the phase transition from the isotropic to the nematic phase using toothpicks. The illustration is educational and it is worth including it as a part of the laboratory work.

**Experimental aims**

- To measure the order parameter for a bundle of toothpicks dropped from different heights.
- To recognize the dependence of the order parameter on the height of fall.
- To recognize similarities with the phase transition.

The equipment used is the same as in the previous experiment. The experiment is very similar to the experiment described in the previous subsection. The only difference is that the experiment has to be repeated several times. Students study the dependence of the order parameter of the toothpick bundle on the height of the drop. As toothpicks dropped from a greater height have more kinetic energy before touching the paper than toothpicks dropped from lower down, the analogy with the internal energy of molecules at higher and lower temperatures is drawn. The toothpicks with the higher speed when they land on the paper rest in a less oriented way than the toothpicks with a lower speed i.e. dropped from the higher or lower height using the same procedure as before.

Students drop the bundle from different heights starting from 3 to 5 cm above the paper and increasing the height in steps of 3 to 5 cm. For each height, the angles are measured and the order parameters are calculated. Finally, the outcomes of the procedure are presented by a graph showing the dependence of the order parameter on the height of the drop. An example of such an experiment is given in figure 4.8.

The results clearly show that above a certain height, which is difficult to determine due to the lack of accuracy of this simple experiment, the order parameter is relatively small and does not depend on the height. For smaller heights, the order parameter increases with decreasing drop height.

If one compares the dependence to actually measured dependences of order parameters or properties directly proportional to them in nematic or smectic liquid crystals on decreasing temperature (figure 1.6(*a*)), and the order parameter's dependence on height (figure 4.8), the similarity is evident. The order parameter's dependence on decreasing temperature is very similar to the order parameter's dependence on the decreasing height of the drop. Actually, the experiment presents a nice mechanical model of a complex phenomenon, just as the phase transition between the orientationally disordered isotropic phase and the orientationally ordered nematic phase is.

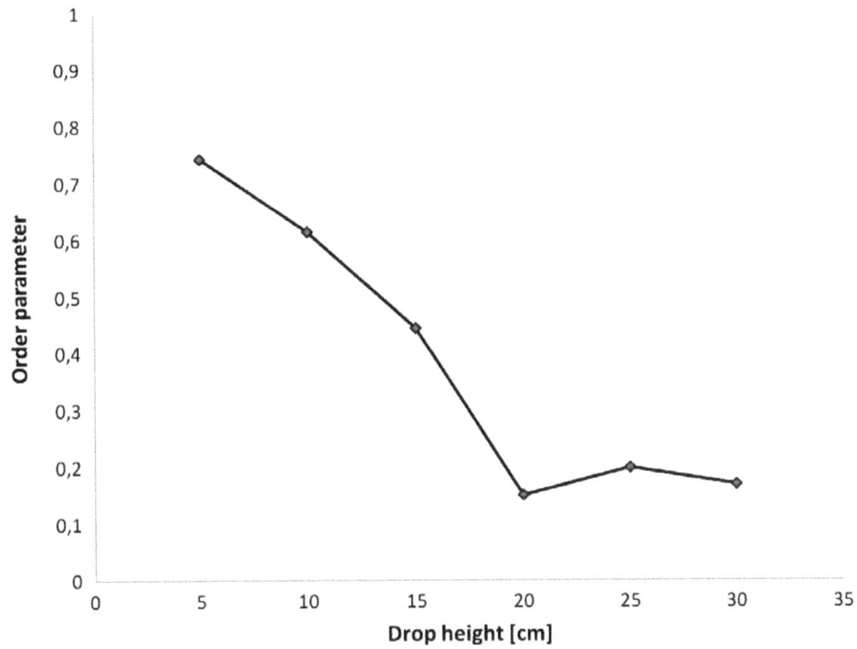


**Figure 4.8.** Order parameter $S_{2D}$ dependence on the height of the drop. The change of order parameter dependence when the height of the drop is around 20 cm is clearly seen. Note that due to the small number of toothpicks, the order parameter never averages to zero.

*Suggestions for the lecturer.* This experiment can be done in the classroom simultaneously, each student experimenting with a different height. If the number of groups is small, the results will give general trends only. On the other hand, the experiment can be done as a hands-on homework. Students should report their measurements electronically and the merged graph should be presented in the next meeting. If so, students may perform measurements for several heights.

As each experimental group performs the experiment with a small number of toothpicks, the effects of the small sample become observable even in the merged results. The value of the order parameter for greater heights is approximately the same and does not depend on the height of the drop. The order parameter is not zero simply because, for a small number of toothpicks, the average is never exactly zero. The consideration could be extended to the question of small volumes, where surfaces are important, because the interactions of molecules among themselves at surfaces are different to when they are in bulk.

**Additional tasks**

1. Study the statistics. Measure the order parameter for a few toothpicks and increase the number of toothpicks in one fall until the measurement of the angles becomes difficult. The height of the drop should remain the same. Draw the dependence of the order parameter on the number of toothpicks.
2. Study the statistics in a different way. Take 20 toothpicks and drop them several times. Measure the angles for each fall and calculate the order

parameter each time. Combine the measurements and draw the dependence of the order parameter on the number of toothpicks included in the calculation. Is the order parameter the same or different to in the previous measurement? What could be the reason for any differences?

3. Use previous measurements for to study the distributions. Compare distributions from the previous two tasks.

## 4.3 Microwave optics of wood

Looking at a piece of wood, one clearly sees the structure of the wood, especially under magnification. Wood fibres are ordered (figure 4.9) in one direction if one does not consider knots. Therefore one can easily expect that properties of wood, such as elastic properties, absorption of water, or even conductivity, is different in the direction along the fibres to in directions perpendicular to it.

But we do not observe only one arrangement in wood. Annual rings, rings of denser structure of wood fibres form circular sheets (*xy* cut in figure 4.9(*a*) and its magnification in (*b*)). Some sorts of wood, oak wood for example, also have significantly different properties in the direction perpendicular to the annual rings (*yz* cut, figure 4.9(*c*)) than parallel to them (*xz* cut, figure 4.9(*d*)). On the other hand, for spruce or pine wood this is not the case.

For our studies, the most important point is that wood is transparent for microwaves. Several typical optical phenomena characteristic for the transmission of polarized light through an anisotropic material can be demonstrated by using wood, for example, anisotropy in absorption and anisotropy in the refractive index. The effects of an anisotropic material on polarized light can be studied in detail.

It seems that there is no important added value for the study of liquid crystals in studying wood in the microwave region. But one can find several analogies. The structure of liquid crystals formed from elongated molecules is analogous to the fibre structure in wood. Additional subtleties in the structure, like annual rings, for example, or knots, change the anisotropic properties of wood. The most important point is that all these structures can be seen by the naked eye, and effects on the transmission of electromagnetic waves that depend on the orientations of the wood fibres or their order, are trivial to measure. The experimental setup is very simple and straightforward. One needs a microwave emitter and a microwave receiver. They are available as undergraduate laboratory equipment. Simple microwave sources have a single frequency and

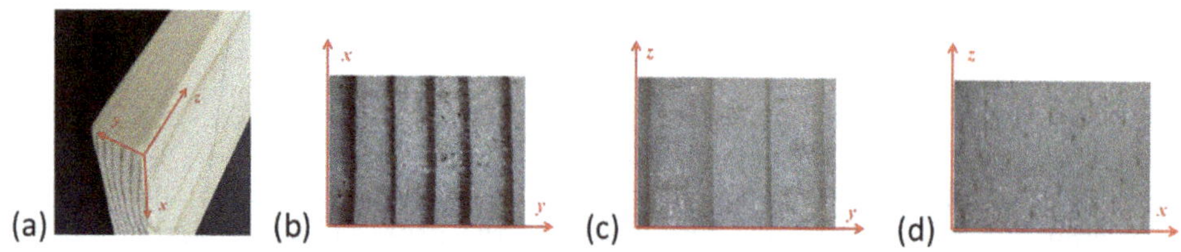


**Figure 4.9.** (*a*) Spruce wood cut in three principal directions: perpendicular to the fibres *xy*, perpendicular to annual rings *yz* and parallel to the annual rings. (*b*) Magnification of wood structure at *xy* cut. (*c*) Magnification of wood structure at *yz* cut. (*d*) Magnification of wood structure at *xz* cut.

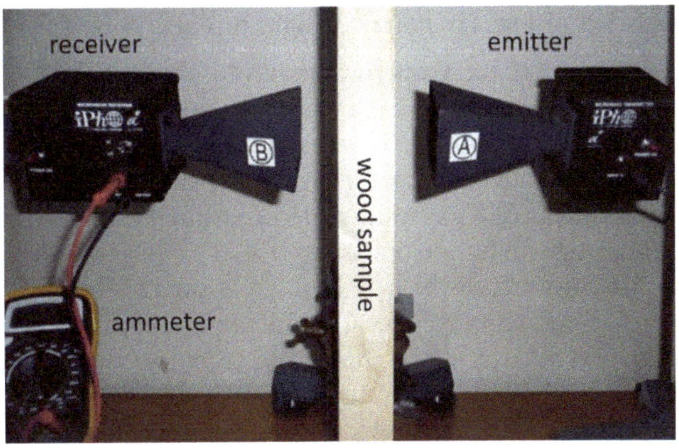


**Figure 4.10.** The microwave setup for measurements of properties of microwaves transmitted through a wood sample.

the polarization of the microwaves is defined by the orientation of the emitting antenna that is fixed with respect to the horn. Therefore microwaves do not need additional polarizers and the direction of the polarization is defined by the orientation of the microwave horn. Similarly, the receiver also detects a single frequency and the component of the polarization parallel to the receiving antenna. Therefore the receiver does not need an additional analyser as the function of an analyser is integrated into the receiver. The sample studied is simply 'illuminated' by microwaves from the emitter and the transmitted microwaves are detected. The detector should be connected to a microphone for demonstration experiments performed in larger lecture halls. For the laboratory, the receiver is connected to the ammeter and the current is measured. The effective current is usually proportional to the amplitude of the measured electric field detected by the receiver. Details about the proportionality, either to the measured amplitude of the electric field or to its square that is proportional to the energy flow, have to be verified for all experimental equipment. More sophisticated microwave sources allow for changes to the frequency. If they are available, dispersion can also be studied in an advanced laboratory, using the same experiments as discussed below. The general setup for these studies is shown in figure 4.10.

Although the optics of anisotropic materials, for example, of liquid crystals, is studied in an earlier section 2.4, the theoretical explanation required in order to understand the experimental results is given within this section. We believe that the interconnection of the visual evidence of changes and the effects on measurements is easier to comprehend through experiments using wood then through experiments that use theoretical expressions to deduce properties that are only indirectly observed. Our suggestion to the lecturer is to allow students to perform a few typical experiments that provide experiences with elliptically polarized electromagnetic waves using wood samples. Next, students should find a setup that allows them to study anisotropy in absorption, and they should also verify whether or not the wood is optically active. All the experiments are described below, however it is suggested that the lecturer leaves some of the studies to students as project tasks.

### 4.3.1 The transmission of linearly polarized light through wood

When an electromagnetic wave, a microwave in our case, enters a transparent anisotropic material, it splits into two waves with perpendicular polarizations, different speeds and, consequently, with different wavelengths. If an incident microwave is polarized and is perpendicular to the surface of a plan-parallel sample (see figure 4.11(*a*)), the linearly polarized microwave generally becomes elliptically polarized after transmission through an anisotropic sample. Let us first consider how to analyse the elliptical polarization state in the case of a wood sample. Let us assume that the typical structure of wood also implies the directions of polarizations where the absorption is the highest and the lowest. If so, a coordinate system that corresponds to the wood sample allows for a straightforward analysis of the state of the transmitted microwave. The electric field for the incident polarized wave is written in the coordinate system where the axis marked as || corresponds to the direction of fibres and the axis marked as ⊥ corresponds to the direction perpendicular to the fibres, as seen in figure 4.11. The coordinate $z$ corresponds to the direction of the wave propagation. The linearly polarized incident microwave is written as

$$\vec{E}_{\text{in}}(t, z) = \left(E_{0\parallel}, E_{0\perp}\right)\cos(\omega t - kz). \tag{4.5}$$

and

$$E_{0\parallel} = E_{\text{in}} \cos\alpha$$

$$E_{0\perp} = E_{\text{in}} \sin\alpha$$

where $\beta$ is the angle between the transmission axis of the polarizer **P** (the direction of the emitter's antenna) and the $y$ axis parallel to the wood fibres (figure 4.11(*b*)). Microwaves split into two waves with polarizations along the fibres and perpendicular to them. The two components have different speeds and during the transmission a phase difference occurs between the two waves. In addition, the energy of the two waves is absorbed differently as an absorption index is a tensorial property as well. The transmitted microwave can therefore be written as

$$\vec{E}_{\text{out}}(t, z) = \left(E_{\parallel}\cos\left(\omega t - kz + \varphi_{\parallel}\right), E_{\perp}\cos\left(\omega t - kz + \varphi_{\perp}\right)\right), \tag{4.6}$$

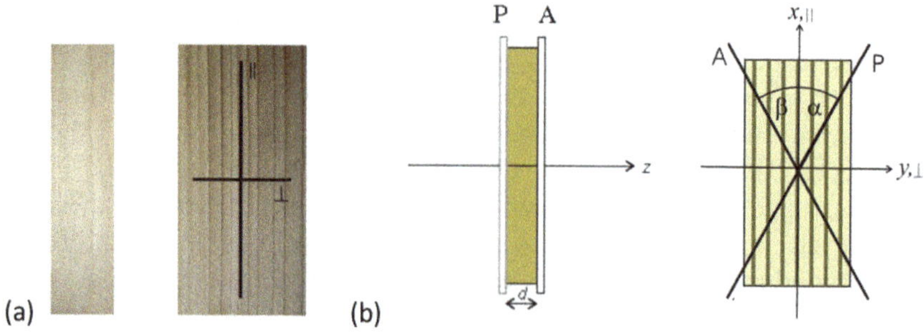


**Figure 4.11.** (*a*) Piece of wood appropriately cut for measurements. (*b*) Experimental geometry for the analysis of measurements used later.

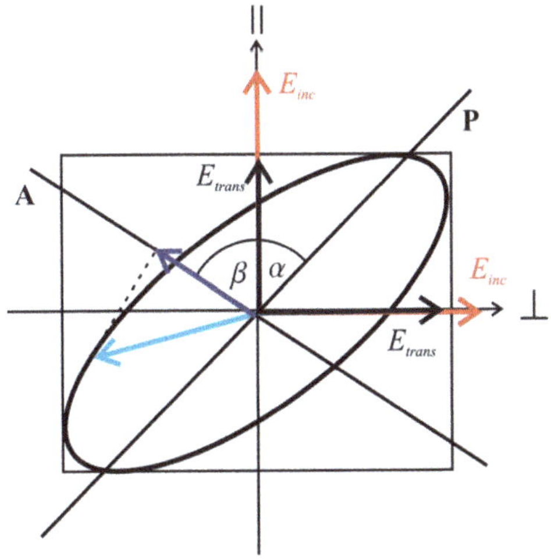


**Figure 4.12.** Characteristic ellipse of transmitted wave and parameters important for the analysis of the transmitted wave.

where $\varphi_\perp$ and $\varphi_\parallel$ are the phase shifts gained by the microwaves of both polarizations while passing through the sample of wood. For the analysis of elliptical polarization the time is renormalized and the elliptical state is given by a phase difference $\delta = \varphi_\perp - \varphi_\parallel$ as

$$\vec{E}_{\text{out}}(t') = \Big(E_\parallel \cos(\omega t'), E_\perp \cos(\omega t' - \delta)\Big). \tag{4.7}$$

Let us visualize the outgoing microwave with respect to the incoming one. Figure 4.12 shows the usual presentation of an elliptically polarized light drawn as an ellipse described by the tip of the electric field vector in the outgoing microwave in one period of time. The polarization of the incident microwave is defined by the orientation of the emitter **P** and the components of the electric field in the incident wave are marked in red. In general, the two beams become phase shifted by $\delta$ and the amplitudes of the electric field are absorbed differently, which additionally influences the shape and the tilt of the ellipse. By studying the intensity of the transmitted microwave for orientations of the polarizer parallel or perpendicular to fibres, one can measure absorption for each polarization separately. However, to measure the difference in refractive indices the whole ellipse has to be known. Let us first describe the measurement of the state of the elliptical polarization. For the derivation of the expressions used later, let us look closer at figure 4.12. The direction of the emitter and the consequent polarization of incident microwaves is marked by **P**. Red arrows are proportional to the incident electric field $E_{\text{inc}}$ decomposed into components parallel and perpendicular to the wood fibres. Partial beams are absorbed differently, as shown by dark arrows that are proportional to the amplitudes of the electric field in the beam after transmission through the wood $E_{\text{trans}}$. The receiver detects only the electric field component parallel to the antenna, marked by A. As an illustration, the light blue arrow shows an instant electric field in a transmitted wave, and a dark blue arrow its component parallel to the receiver's antenna. Finally, the electric field component of the transmitted wave along the receiver's antenna oscillates with the amplitude marked as a thin dashed line.

The receiver detects only a component of the electric field along the antenna in direction $\vec{e}_A = (\cos\beta, \sin\beta)$.

$$\vec{E}_{\text{out},A} = \left(\vec{E}_{\text{out}} \cdot \vec{e}_A\right)\vec{e}_A \tag{4.8}$$

The intensity of the transmitted microwave in a time average of the detected signal over the period is

$$\begin{aligned} I &= \frac{1}{2}\varepsilon_0 c \bar{E}^2_{\text{out},A} \\ &= \frac{1}{2}\varepsilon_0 c E_\perp^2 \cos^2\alpha \cos^2\beta + E_\parallel^2 \sin^2\alpha \sin^2\beta \\ &- \frac{1}{2} E_\perp E_\parallel \sin 2\alpha \sin 2\beta \cos\delta, \end{aligned} \tag{4.9}$$

where $\alpha$ and $\beta$ are defined in figure 4.12. The rotation of the polarizer and the analyser, that is the horns of the emitter and the receiver, allow measurements of various properties of the elliptically polarized wave to be taken. If the principal axes of the sample are known, a simple variation of the setup and two different measurements allow the extraction of absorption coefficients and the phase difference $\delta$. Measurements using wood allow students to see the connection between the visually observable wood fibres' direction and principal directions. However, measurements of the anisotropic properties of crystals and liquid crystals in optical regions allow students to deduce principal directions and the birefringence from the results of optical observations, but they see no direct visual evidence with respect to the underlying structure. Therefore, experiments with wood provide students with several very important experiences about the effects of anisotropic structures on the transmission of light.

### 4.3.2 Familiarization with the measurements

Let us first conduct a few experiments with microwaves in order to familiarize students with the equipment. The first two experiments come from the usual arsenal of experiments with microwaves, the rest provide students with experiences of microwave transmission through an anisotropic material, using wood as an example. In later subsections, experiments allowing more detailed measurements of anisotropic properties are discussed, but this section should be considered as preparation for them. Because all the experiments suggested here start from, essentially, a very similar setup, this subsection is organized in a different way to the others. First the requisites are listed. The aims for the separate experiments are then given and the corresponding experiments discussed.

**What is required for the experiments**

- A microwave set with a transmitter and a receiver.
- An ammeter for measuring the received signal.

- A metallic plate.
- A piece of glass.
- Pieces of styrofoam of different thicknesses.
- Wire net with metal wires in one direction only. The metal wires should be approximately 5 mm to 10 mm apart.
- Pieces of different wood of approximately the same thickness, cut as in figure 4.11(*a*).
- Pieces of the the same material of at least two different thickness, cut as in figure 4.11(*a*).

The setup for microwave studies:

**Experimental aims**

- To build a setup for taking measurements.
- To become familiar with the setup.
- To find the best positions for the transmitter and receiver ready for later studies.

The microwave transmitter and receiver are assembled according to the manual. Horns should be rotatable around their long symmetry axis. Usually the angles of rotation are marked. For demonstration purposes, it is useful to mark the orientation of the antenna on the outer part of the horn by colouring or with an additional double arrow. The magnitude of the received signal is measured. Some kits are already equipped for the measurement of the received signal. If not, an ammeter should be connected directly to the receiver. The signal is proportional either to the amplitude of the electric field or to its square, depending on the equipment. An assembled experimental setup looks like the setup in figure 4.10 or 4.13. The space between the horns is used for inserting various samples like the wood sample in figure 4.10 or studies can be done with a free space, as will be described later.

Several observations that lead to more detailed experiments later are useful in the 'familiarization' period. First, the range for measured current is determined by rotating the receiver's horn until the maximal signal is obtained. Next, the dependence of the detected signal on the distance between the horns is studied. As standing waves can occur, the intensity of microwaves does not simply decrease with increasing distance between the emitter and receiver as one would expect. For later experiments, the receiver should be positioned where the signal is clear and strong but where it is not in a shoulder of a standing wave. The distance between horns is usually chosen to be shorter than the first maximum.

A piece of thick glass is put between the horns and the effect on the signal is observed. Then pieces of styrofoam are put between the horns. Although styrofoam is not transparent for visual light it is transparent for microwaves. A thickness-dependent absorption is verified by adding more layers of a styrofoam. Finally, a metallic plate is put between the horns. Metal is not transparent to microwaves.

*Tricks of the trade.* The time students spend playing with a microwave setup is very valuable. There are several phenomena that can be studied separately, but for later experiments one has to take care that these phenomena do not play a significant role. For larger distances between horns, standing waves can occur and they are easily detected. However, when microwaves pass through the transparent material, the wavelength changes and the condition for a standing wave can occur although it was initially avoided. For larger distances between horns, reflections from nearby surfaces often occur and may influence the measurements. Reflection also occurs on the observed materials. For thicker samples, one can come across the situation where the effects known from visual optics as 'thin layer interferences' can be observed.

All of these problems can be avoided by the proper positioning of the horns, however, if the distance between horns is given in advance, students would not think about the possible situations they may find later in research.

*Suggestions for the lecturer.* Old microwaves emitters usually came with horns that were rotatable. New versions of microwave sources are often pre-designed for experiments suggested in associated accompanying materials. One microwave source which has recently become available comes hidden within a box and no rotations, except the rotation of the box as a whole, are possible. We strongly advise not to use such black box experimental equipment, if possible; find microwave sources where the antenna and horns are visible and rotations are supported.

**Experimental aims**

- To find the relative orientations of both horns where the received signal is minimal.
- To find the relative orientations of both horns where the received signal is maximal.
- To become acquainted with Mallus' law for microwaves.

The rotation of horns has the same effect in experiments using electromagnetic waves as in the visual wavelengths' range rotations of the polarizer or the analyser. However, the experiences of students coming across microwaves for the first time are different. The intensity of an emitted microwave is observed through an ammeter and is observed indirectly. However, the polarization of the microwave is determined by the orientation of the antenna fixed in a horn. This visual evidence of the determination of polarization helps students comprehend the concept of polarization.

Two special relative orientations are of the utmost importance. The horn of the emitter is fixed and the receiver's horn is rotated. The reading of the ammeter increases and decreases upon rotation. One orientation exists where the orientation of the receiver's antenna (the orientation of the horn) coincides with the orientation of the transmitter's antenna. For this special relative orientation the received signal is maximal. However, the signal is not detected if both antennas are perpendicular to each other.

Malus' law says that the amplitude of the electric field in the wave transmitted through the analyser or the amplitude of the signal that is detected by a receiver is proportional to the projections of the electric field in the incident wave. Using the geometry given in figure 4.11, the amplitude of the electric field and the corresponding intensity of the transmitted wave is

$$E_{0,\mathrm{rec}} = E_{0,\mathrm{inc}} \cos(\alpha + \beta)$$
$$I = \frac{1}{2} E_{0,\mathrm{inc}}^2 \cos^2(\alpha + \beta) \tag{4.10}$$

where zeros mark the amplitudes of the electric field in the electromagnetic wave. The sum $\alpha + \beta$ is the angle between the transmitting and the receiving antennas. A detailed measurement shows that rotation of the receiving horn by 180° gives all the information about any signal. Detailed measurements depend on the characteristics of the equipment. If the receiver's signal is proportional to the amplitude of the detected electric field, the signal will approach zero for crossed antennas linearly. If the intensity of the signal $I$ is detected, the measurement approaches zero quadratically.

**Experimental aims**

- To study anisotropy in absorption for a simple metallic wire net.
- To verify that the wire net acts like a polarizer.

In this experiment, students are introduced to the concept of anisotropy in absorption. The aim is to observe and measure the intensity of transmitted microwaves through the wire net consisting of parallel metal wires. The transmitter and the receiver antennas are positioned in orientations where the detected signal is the largest, that is, parallel. The wire net is placed between the two horns instead of the wood in figure 4.10, and it is slowly rotated around the normal to the net. Two orientations of the wire net are special. For one orientation the signal cannot be detected at all, for another the signal is almost the same as without the net.

This simple result is very similar to that obtained for the detection of the signal dependent on the orientation of the receiving antenna (horn). The signal is absorbed if the wires are parallel to the polarization direction. The wire net does not affect the signal if the wires are perpendicular to the polarization direction of the microwaves. The wire net has therefore been shown to have anisotropic properties. The absorption coefficient for polarization along the wires is very large, but for polarization perpendicular to the wires, it is small. This is a typical property of a polarizer but also a typical anisotropic property, as the results of the measurement depend on the direction of the material with respect to an internal stimulus.

However, if the transmitted signal is measured in detail, several other subtleties are revealed. Let us assume the coordinate system has one axis ($x$) along the wires and the other one ($y$) perpendicular to them. The angles $\alpha$ and $\beta$ present the angles that are formed by the polarizer and the analyser with the $y$ axis, respectively. The amplitude of the electric field in the transmitted microwave is

$$E_{\mathrm{tr}} = E_0 \cos \alpha \tag{4.11}$$

but the receiving antenna detects only the component along the direction given by $\beta$.

$$E_{\mathrm{det}} = E_0 \cos \alpha \cos \beta \tag{4.12}$$

If antennas are parallel, $\alpha = \beta$ and the detected signal is

$$E_{\text{det}} = E_0 \cos^2 \alpha \tag{4.13}$$

which is quadratic and dependent on $\alpha$. However, to verify that the wire net behaves like a polarizer, an additional measurement is needed, as a similar result could be obtained for a non-absorptive material as well. The wire net is placed at a general angle $\alpha$ and the receiving antenna is rotated from $\beta = -90°$ to $\beta = +90°$. For any orientation $\alpha$, the detected signal should vanish for $\beta = \pm 90°$. Another verification is that one keeps the relative angle between the receiving antenna and the direction of the wires fixed at 90° and the antenna rotates parallel to the wires. The signal should remain zero.

*Suggestions for the lecturer.* The polarized electromagnetic wave is usually presented as a wave where the spatially changing electric field appears to be similar to a mechanical transversal wave. This often gives students an erroneous impression that the polarization of an electromagnetic wave is very similar to a mechanical transversal wave so they then expect that the signal with polarization parallel to the wires is transmitted and the signal with polarization perpendicular to them is not. The lecturer should attempt to eliminate this false perception.

- To observe that the effect of the wood on the microwaves is dependent on the orientation of the wood.
- To become familiar with the most efficient setup for measuring the absorption.
- To become familiar with the most efficient setup for measuring the birefringence.

Horns are oriented perpendicularly to each other in orientations with a minimal detected signal, in the crossed setup. A piece of wood is inserted between the horns as in figure 4.10. The wood is slowly rotated between the horns around the normal to the largest side of the sample. The signal increases and decreases depending on the orientation of the fibres with respect to both antennas. It becomes evident that for fibres parallel or perpendicular to either the emitter's or the receiver's antenna, the signal is not detectable. For other orientations, the inserted wood is also the reason that the signal is detected in the crossed setup. It is interesting to discover the orientation of the wood required for a maximal signal. The experiment is repeated with different sorts of wood and with one sort of wood of different thicknesses. The amplitude of the measured signal and the orientation of the wood for the maximal detected signal vary with the sort of wood and with the thickness of the sample.

*Tricks of the trade.* For demonstration, the piece of wood is simply held with a hand and rotated.

For more detailed measurement, the piece of wood should be fixed and both horns rotated by the same angle, 5° or 10° in one step. This procedure will be used later.

All samples used in the experiments need to be verified previously. For thicker samples, rotations of multiples of $\pi$, or close to that value, may occur and if so, the measurement of the signal variation upon wood orientation can be obscured.

Humidity also severely influences measurements. If the wood sample is not very dry, the absorption of microwaves can be very high. Therefore use very dry wood samples only. Even so, as humidity usually depends upon the season, for example, during the heating period it is much lower, results also depend on the humidity of the air as that influences the properties of the wood samples. Therefore rather large variations in results are expected and measurements by different students of absorption coefficients and birefringence of the same sample at different times may significantly vary, but they should not be considered incorrect.

*Suggestions for the lecturer*. As a semi-quantitative measure of the intensity of a transmitted wave, when experiments are performed in a lecture room a sound signal could be used. Besides the well known fact that using more senses to interpret a single result helps learning, the sound signal is also impressive and motivating.

**Additional tasks**

1. Play with your microwave setup. Find the proper range of the ammeter to cover the whole range of signals. Rotate both horns independently. Study the dependence of the signal on the distance between the horns for parallel antennas. Find the distance where the setup is the most sensitive to the rotations of the horns.
2. Measure the dependence of the detected signal on the rotation of the receiver's horn.
3. The metal plate not only absorbs microwaves but also reflects the microwaves. Construct the setup for the observation of reflected microwaves. Verify the polarization of the reflected microwaves and measure its amplitude.
4. Also, the wire net reflects microwaves. Verify the polarization of reflected microwaves. What is the direction of polarization with respect to the polarization of incident microwaves?
5. Study the receiver's signal for general angles $\alpha$ and $\beta$ with respect to the wires in the net.

### 4.3.3 Refractive indices

Refractive indices of transparent materials are determined by measuring the incident and refraction angles for an oblique incidence of light. As the trajectory of light cannot usually be followed in pure transparent samples due to the lack of scattering, refraction angles are deduced from the direction of light in the air after refraction. To enhance the effect of the refraction of light, material in the form of a prism serves well, as already seen in section 2.4. Let us use the same principle and measure the refractive indices for polarizations perpendicular to and parallel to the fibres. The idea is simple. The microwave beam passing the wooden prism is deflected. The angle of deflection allows the refractive index to be calculated.

This measurement is important if the lecturer wants to compare a direction dependent absorption coefficient to a theoretical expression. The absorption coefficient in a uniaxial material depends on principal refractive indices and on principal absorption coefficients.

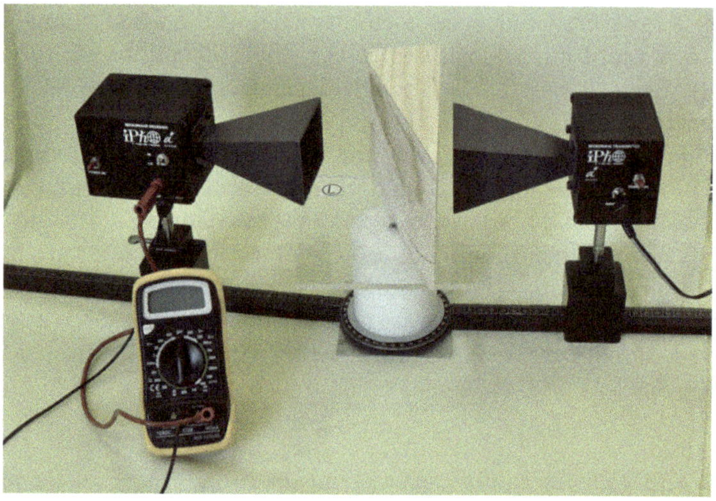

**Figure 4.13.** The setup for the measurement of the principal refraction coefficient. The receiver is mounted on a goniometer allowing the angle dependent amplitude of the received microwave signal to be studied.

**Experimental aims**

- To measure the refractive indices of wood for polarizations of microwaves perpendicular to and parallel to wood fibres.

**What is required for the experiments**

- A wooden prism.
- A microwave emitter and a receiver on a goniometer.
- An ammeter for measuring the received signal.
- Support for a wooden prism.

The setup is seen in figure 4.13. The geometry of the measurement is very simple if the incident microwave beam is perpendicular to one side of the prism. To guarantee the relative orientation of the beam polarization with respect to the wood fibres, one side of the wooden prism is cut along the wood fibres. This side corresponds to the side of the microwave incidence. The profile of the microwave beam is first measured without a prism. The receiver is set on the goniometer and the intensity of the microwave is measured independent of direction. The microwave beam is rather wide and a typical angular profile for a detected amplitude is seen in figure 4.14(*a*).

After the profile of the microwave beam has been measured, the wooden prism is placed on the stand in such an orientation that the incident beam is perpendicular to the side that is cut parallel to the fibres. The polarization of the incident beam is either parallel to or perpendicular to the wood fibres. Again, the angular profile of the transmitted microwave beam is measured. The maximum in the profile is found at an angle indicating that the whole beam is deflected due to the refraction of the microwaves in the wood. The measurement for the prism shown in figure 4.13 is presented in figure 4.14. The refractive index is calculated from the refraction angle

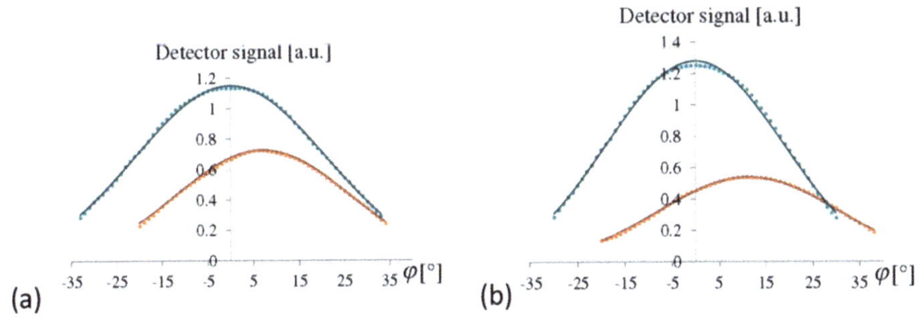


**Figure 4.14.** (*a*) The angular profile of a direct microwave beam (blue) and the angular profile of a refracted beam polarized perpendicular to the wood fibres (red). (*b*) The angular profile of a refracted beam polarized parallel to wood fibres. The received signal is given in arbitrary units. Note that for both refracted beams, maxima of the signal are shifted from the central position for the non-refracted signal.

at the wood–air interface when the wave is leaving the wood. The incident angle is then equal to the apex angle of a prism $\gamma$, and the refraction angle $\beta + \gamma$ is measured from the position of the profiles maximum. The geometry for the calculation is exactly the same as the geometry for the wedge cell with a liquid crystal, see figure 2.15, with an incident angle $\alpha$ equal to zero.

$$n_{\parallel,\perp} = \frac{\sin\left(\beta_{\parallel,\perp} + \gamma\right)}{\sin \gamma}. \tag{4.14}$$

Here, subscripts $\parallel$, $\perp$ stand for polarization parallel or perpendicular to the wood fibres and the corresponding principal refractive indices.

The detected profiles are significantly different and allow maxima to be determined in both cases. The maximum determines the deflected beam direction of propagation. The profile of the refracted beams is seen in figure 4.14.

The microwave beam is wide if compared with laser beams. One should be aware that laser beams compared with the wavelength of the light are much wider than the microwave beam in comparison to the microwave wavelength. As the dimensions of a microwave beam are comparable to the dimensions of the experimental equipment, the upper part of the microwave beam profile has to be fitted with a bell shaped curve and the direction for which the intensity is maximal is extracted from the curve.

*Tricks of the trade.* The same wedge-shaped wood can be used for the determination of absorption coefficients for both polarizations. If the lecturer plans a laboratory with precise measurements of anisotropic properties in the microwave region, the wedge shaped pieces of different sorts of wood can be prepared in advance. The wood should be dry even before cutting, but samples also have to be stored in a dry place when they are not being used. It is useful for the humidity to be registered when the measurements are made. If different students measure the same samples at different times, humidity may have a detectable impact on the results.

*Suggestions for the lecturer.* The same experiment can be used to measure the refractive index for other isotropic materials which are transparent for microwaves. Examples are various styrofoams, wax samples or other plastics.

**Additional tasks**

1. In the presented experiment the incident beam is perpendicular to the side which has been cut parallel to the wood fibres. Discuss the differences between this experiment and the experiment where the incident beam is perpendicular to the opposite side of the wooden prism, as in the original experiment. Fibres are not parallel to the surface on that side. Try to measure the difference. Is it observable? If so, which polarization is more sensitive to the change of the incident surface?
2. Discuss the differences if the incident beam is oblique. Measure the signal for oblique incidence.
3. Perform the experiments discussed in the first two questions, deduce the refractive indices and compare them to the principle values measured in the original experiment.

### 4.3.4 Anisotropy in absorption

Liquid crystals are always studied in relatively thin cells. The absorption is not considered an issue. But in the newest applications, where liquid crystals are used for waveguides, absorption is the phenomenon that becomes important and has to be considered in detail. As liquid crystals in liquid crystalline phases are anisotropic, anisotropy in absorption is also expected. The manifestation of anisotropy in absorption and its consequences can be studied using properly cut wood and microwaves [2].

Absorption in a material is described by a spatial dependence of amplitude of the propagated microwave.

$$I = I_0 \exp(-\lambda x). \tag{4.15}$$

Here $I$ is the intensity of the beam at the distance $x$ from the material surface, $I_0$ is the intensity of the incident beam and $\lambda$ is the absorption coefficient. In anisotropic wood $\lambda$ depends on the direction of wood fibres and values for polarizations parallel and perpendicular to fibres are usually given.

**Experimental aims**

- To measure the absorption coefficient for polarizations parallel to and perpendicular to the wood fibres.
- To measure the absorption coefficient for a general orientation of wood fibres.
- To compare the measured dependence of the absorption coefficient on the direction of microwave propagation with theoretical expressions.

**What is required for the experiments**

- Two wedge-shaped pieces of the same type of dry spruce or pine wood (see figure 4.15(*a*)).
- Clamps for the wedge-shaped wood pieces.
- Pieces of wood, of the same thickness, cut in different directions with respect to the fibres (see figure 4.15(*b*)).

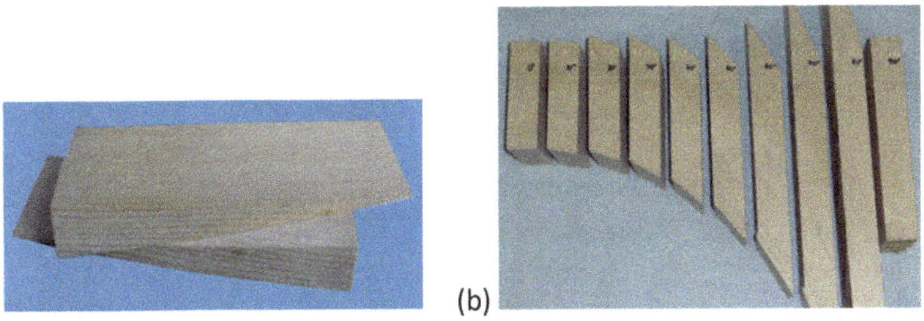


**Figure 4.15.** (*a*) Two wedge-shaped pieces of wood are combined to allow the continuous variation of sample thickness (Photo: Saša Ziherl). (*b*) Pieces of spruce wood of the same thickness, cut in different directions with respect to wood fibres (Photo: Saša Ziherl).

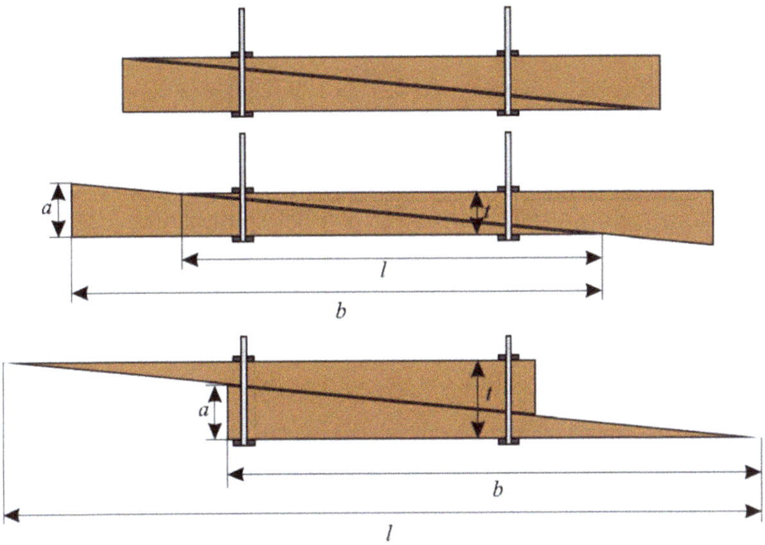


**Figure 4.16.** Wedge-shaped pieces of wood allow continuous variation of the sample thickness, from much less then the shortest side of the wedge to almost twice as thick.

- A microwave emitter and receiver.
- An ammemter for measuring the received signal.
- Holders for the measured pieces of wood.

The piece of dry spruce wood, of dimensions approximately 10 cm thick, 10–15 cm wide and about 50 cm long, should have its fibres along the longest dimension and should be without knots. The piece of wood is cut along the diagonal, see figure 4.15(*a*). With the two pieces of wood fixed together as shown in figure 4.16, the thickness of wood can be varied easily between 5 cm and 15 cm.

The emitter and the receiver are positioned facing one another. The receiver's orientation is parallel to the polarization of microwaves. Both the wedge-shaped pieces of wood are fixed together between the emitter and the receiver, as seen in

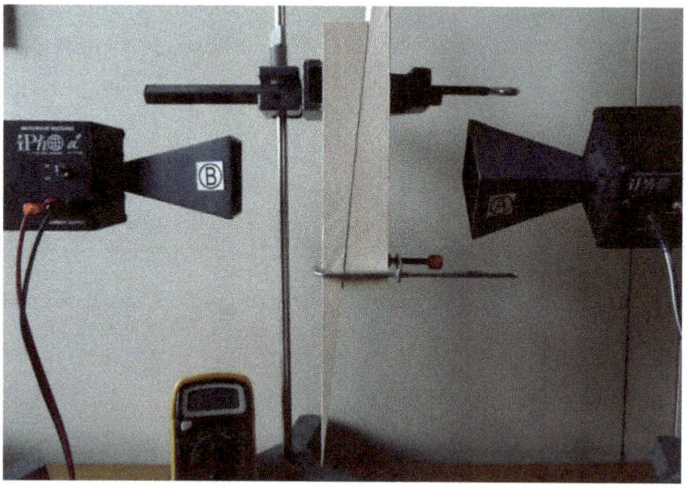

**Figure 4.17.** The setup for the measurement of anisotropy in absorption for wood (Photo: Saša Ziherl).

figures 4.16 and 4.17. The transmitted signal is measured twice for each thickness of wood. The emitter and the receiver are set parallel to the fibres first and then perpendicular to the fibres. Each measurement is part of a different set of measurements. The first measurement is part of the set for the determination of the absorption for polarization parallel to the fibres, and the second is for the set for the determination of the absorption for polarization perpendicular to the fibres. The thickness of wood is gradually changed by shifting the two wedges, one with respect to the other, as indicated in figure 4.16.

Coefficients $\lambda_{\parallel}$ and $\lambda_{\perp}$ for polarizations parallel to and perpendicular to the fibres, respectively, are extracted from the linear relation

$$\log\frac{I}{I_0} = -\lambda x \tag{4.16}$$

If wedge-shaped wood samples are not available, a similar experiment can be done by consecutive measurements of one, two, and three pieces of the same sort of wood put together. The fundamental idea about anisotropy in absorption can be recognized even with a single piece of wood. Although a reflection does not allow for an accurate measurement of absorption coefficients, the semi-quantitative difference, that is which polarization is more absorbed and which is less, is still clearly detectable.

When a material is anisotropic, it is usually anisotropic in any property that is related to a vectorial stimulus. As the structure of wood is clearly visible, the expected differences in properties can be easily associated with the structure. For example, one expects a larger refraction coefficient for polarization along the wood fibres because the induced polarization along fibres is expected to be larger then that perpendicularly to them. Also, the absorption along fibres is expected to be larger due to the orientation of vessels in the wood. The absorption of water or oil also depends on the orientation of wood fibres, and the elastic properties in different directions are different too.

Absorption and refraction are two typical phenomena that reflect anisotropic properties. Both are also direction dependent. The analysis of direction dependent properties is especially easy with wood samples as wood can be cut in any direction. The analysis is much simpler if the incident beam is perpendicular to the surface and no refraction in the direction of the wave vector occurs. General directions are usually obtained by oblique incidence, which complicates the measurements and the analysis. Therefore, for the second experiment, samples of wood of the same thickness are cut in different directions with respect to the wood fibres. Spruce wood serves very well for this purpose.

The measurement is not very detailed as a large wedge-shaped sample of wood cut in a general direction with respect to the wood fibres is almost impossible to get. The original timber should be very large and homogeneously structured. But samples of one or two thicknesses for each direction of fibres (figure 4.15(*b*)) is not so difficult to obtain. The measurement of both absorption indices for one thickness provides only an estimation. If two thicknesses are available, one can measure each sample separately and the combination of the two, which provides data for three thicknesses. Anyway, even a measurement of a single sample provides enough data to follow the general dependence of the absorption coefficient on the propagation direction of microwaves with respect to the fibres.

*Tricks of the trade*. The choice of wood should be based on the uniaxiality or biaxiality of the wood. If the wood can be described by three different refractive indices, one for polarization parallel to the fibres, one for polarization perpendicular to the fibres and perpendicular to the annual rings and one for the polarization perpendicular to fibres and parallel to the annual rings, the analysis becomes more complicated. Such wood is biaxial. An example of such wood is oak. In order to tell whether a piece of wood is uniaxial or biaxial, it should be cut, perpendicularly to the wood fibres, from a large log which has annual rings that are quasi-straight. If the wood is biaxial, the birefringence, that is a difference between refractive indices for polarizations parallel and perpendicular to the annual rings, is significant.

*Suggestions for the lecturer*. The presented experiments are of an educational nature. They provide a direct, visually evident link between the structure and the phenomena that are related to the anisotropy of the structure. In studies of liquid crystals, it is the opposite. The transmitted light is detected and the structure is deduced from measurements. If students have experienced this link through studies of wood, it is much easier for them to consider possible structures behind the signals than to construct the structures from equations discussed only theoretically.

**Additional tasks**

1. Measure absorption coefficients of samples cut from different sorts of wood and try to relate the results to the structure.
2. Compare measurements of the same samples performed on days with different air humidities. How does absorption depend on humidity?

3. Measure the absorption of fresh samples and of older samples of the same sort of wood. How does the absorption depend on the dryness of the wood? Note that similar measurements are used for the control of wood freshness.

### 4.3.5 Birefringence

Let us start with the theoretical explanation. We study in more details a uniaxial wood sample such as spruce or pine wood. Therefore the properties of wood are known if we know the values for the refractive index and absorption coefficient for polarization perpendicular to wood fibres, $n_o$ and $\lambda_o$, respectively, and for the refractive index and absorption coefficient for polarization with a component along the fibres, $n_e$ and $\lambda_e$. In uniaxial anisotropic materials both $n_e$ and $\lambda_e$ depend on the direction of propagation and are given as

$$\frac{1}{n_e^2(\theta)} = \frac{\cos^2\theta}{n_o^2} + \frac{\sin^2\theta}{n_e^2}, \tag{4.17}$$

$$\lambda_e = \frac{\lambda_o n_e^4 \cos^2\theta + \lambda_e n_o^4 \sin^2\theta}{\left(n_e^2 \cos^2\theta + n_o^2 \sin^2\theta\right)^2}. \tag{4.18}$$

Here $\theta$ is the angle between the optical axis (along the fibres) and the wave vector $k$ of the propagating wave. To measure the difference between refractive indices, one has to know the absorption coefficients. The characteristic ellipse does not depend on the phase difference between the two perpendicularly polarized waves only, but also rotates due to more intensive absorption of one polarization with respect to the other.

When the absorption of a wood sample is known, the difference in refractive indices between the two perpendicularly polarized microwave beams can be measured. The characteristic ellipse describing the elliptically polarized wave can be deduced from the measurement of the transmitted signal for different orientations of the receiver. As absorption is present, it has to be included in components of the transmitted wave. Therefore the long axis of the ellipse may rotate not only due to the phase difference $\delta$ but also due to absorption. Equation (4.19) is for the intensity of the transmitted signal and is the same as for light so it is repeated here.

$$\begin{aligned} I &= \frac{1}{2}\varepsilon_0 c\bar{E}_{\mathrm{out},A}^2 \\ &= \frac{1}{2}\varepsilon_0 c E_{\mathrm{out},\perp}^2 \cos^2\alpha\cos^2\beta + E_{\mathrm{out},\|}^2 \sin^2\alpha\sin^2\beta \\ &- \frac{1}{2}E_{\mathrm{out},\perp}E_{\mathrm{out},\|}\sin 2\alpha\sin 2\beta\cos\delta. \end{aligned} \tag{4.19}$$

However, absorption in wood is not negligible and the amplitudes of the electric field decrease exponentially. For amplitudes of electric field in the transmitted signal the absorption index is inserted:

$$
\begin{aligned}
E_{\text{out},\|} &= E_{\text{in},\|} \exp\tfrac{1}{2}\left(\lambda_e(\theta)qd\right) \\
E_{\text{out},\perp} &= E_{\text{in},\perp} \exp\tfrac{1}{2}\left(\lambda_o d\right)
\end{aligned}
\tag{4.20}
$$

Here $\lambda_e(\theta)$ gives the direction dependent absorption index, as given in (4.18). If the receiver detects the amplitude of an electric field, the square root of the expression given in equation (4.19) has to be used for the comparison of measurements.

The last set of experiments illustrates the rather complicated dependencies given in the equations above. In my opinion, it is the most demanding of all the experiments with wood. Determination of the birefringence requires a set of measurements for one sample only. For more accurate measurement, the amplitude or the intensity of the transmitted microwaves have to be measured for at least a few different thicknesses. Therefore these experiments demand an extensive time slot for several measurements, computer skills for data extraction and also a good understanding of the underlying phenomena. The experiments in all their complexity can be left for project work, but some of them should at least be shown as demonstrations.

**Experimental aims**

- To measure the birefringence of various sorts of wood.
- To measure the dependence of the refractive indices on the direction of microwave propagation with respect to the wood fibres.
- To compare the difference between the measured dependence of two refractive indices on the direction of microwave propagation with respect to the fibres, with the theoretical expression.

**What is required for the experiments**

- Pieces of various wood samples.
- Pieces of wood of the same thickness cut in different directions with respect to the fibres (see figure 4.15(*b*)).
- A microwave emitter and receiver.
- An ammeter for measuring the received signal.
- Holders for the measured pieces of wood.

First, the two principal absorptions are measured for the piece of wood being studied. The polarization of incident microwaves to the wood is parallel to the fibres and the receiver is rotated so its antenna is also parallel to the polarization of the microwaves and the wood fibres. The transmitted signal is measured. The horns are then rotated by 90° and the transmission is measured again, now for polarization perpendicular to the fibres. The wood is removed and the signal without the wood is measured again. The estimation of the absorption coefficients is calculated using equation (4.16).

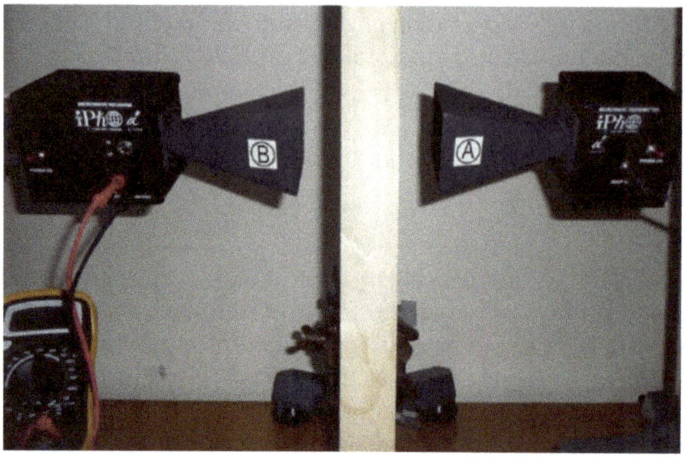


**Figure 4.18.** The setup for the measurement of birefringence. The emitter is marked A, the receiver is marked B, and both horns can be rotated. Note that the emitter horn is at a general angle (the best choice is 45°) with respect to the wood fibres. Only the receiver's horn is rotated during the experiment *in praxis*.

The setup for the measurement is as follows: the piece of wood is fixed between the emitter and the receiver, see figure 4.18. The polarization of the emitted wave is set to 45° with respect to the wood fibres. The receiver's polarization is initially parallel to that of the emitter. The receivers horn is rotated stepwise and the signal is measured. The measured intensity or the amplitude of the received signal, depending on the equipment, is analysed according to the expression given in equations (4.19) and (4.20). For absorption coefficients estimated values are taken from the first experiment in this set. The phase difference $\delta$ is extracted from the measurement by fitting the measured curve. From the phase difference $\delta$, along with both absorption coefficients and the microwave wavelength, the birefringence $\Delta n$ can be calculated for a known thickness of wood.

$$\Delta n = \frac{1}{2\pi}\frac{\lambda\delta}{d} \tag{4.21}$$

The measurement is more detailed if the wedge-shaped piece of wood described in the previous subsection is used. The thickness of the sample is controlled, the phase difference $\delta$ is measured for several thicknesses and the linear dependence of $\delta$ with respect to the wood thickness allows for a more accurate estimation of the birefringence. An example of such a measurement is seen in figure 4.19.

The same procedure is carried out for wood samples made of different sorts of wood, and for pieces of wood cut in various directions with respect to the wood fibres. In addition, splinter board is measured and the properties of isotropic materials are illustrated.

The last set of experiments connect several phenomena. The elliptical state of transmitted microwaves is determined by anisotropy in the refractive index as well as by anisotropy in absorption. Both effects are important. Different properties are expected for stimuli parallel to the fibres and perpendicular to it. A dependence of

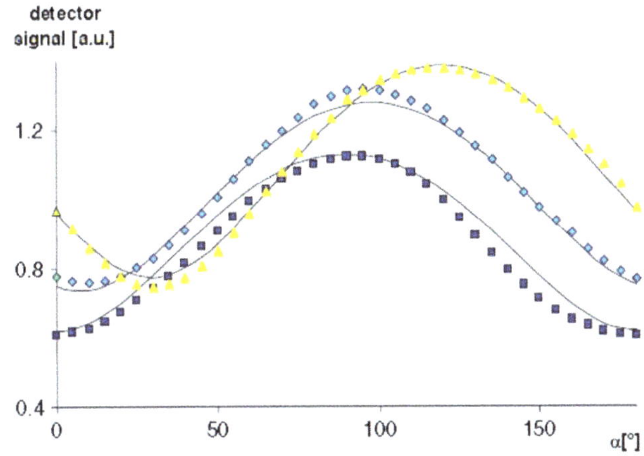


**Figure 4.19.** Measurements of the intensity of a transmitted microwave signal through samples of spruce wood of thickness 5.3 cm (squares), 4.6 cm (diamonds) and 3.4 cm (triangles). The incident microwaves are polarized under 45° with respect to the wood fibres. Lines give fits according to equations (4.19) and (4.20), after [7]. Note that fits for different thicknesses have the same absorption coefficients and birefringence.

one of the indices and one of the absorption coefficients could be understood by a hand-waving geometrical argument. However, the set of experiments evidently shows the effects.

*Tricks of the trade.* Samples should not be too thick. For thicker samples, interference effects of reflected waves can cause problems. Similarly, it should be verified that the combination of two wedges does not become to thick, otherwise the interference has to be considered in the analysis, which make results less evident.

Some sorts of wood are strongly biaxial, for example, oak. For microwave propagations perpendicular to the fibres, the results are the same. Additionally, the difference between two smaller refractive indices can be shown. The sample, cut perpendicularly to the fibres, has two distinctive directions, perpendicular to and parallel to the annual rings. Also, for directions of polarizations parallel to these two directions, the refractive indices are different and the birefringence can be measured. Studying samples cut in general directions can offer a plethora of new phenomena, which are more complicated to analyse and allow students to gain a deep insight into the optics of anisotropic materials.

*Suggestions for the lecturer.* The experiments presented here can be used to demonstrate the phenomena discussed. When the emitter's antenna and the receiver's antenna are set perpendicular to each other, no signal is detected. Inserting a piece of wood between them, with the wood fibres oriented under 45° with respect to both antennas, the signal is suddenly detected. Rotating the wood, two orientations are found, parallel and perpendicular to the wood fibres, where the signal is not transmitted. Both orientations correspond to principal directions of polarization.

Measuring several pieces of wood is tedious and boring. If several microwave sources are available, different students should measure different sorts of wood and different orientations of wood, at the same time. Even if only a few microwave sources are available, different students should measure different samples, but they

**Table 4.2.** Birefringence data for different sorts of wood. The same material was measured in 2008 and 2013. Humidity, age and other properties influence the birefringence [7].

| Sort of wood | $\Delta n$ (2008) | $\Delta n$ (2013) |
|---|---|---|
| Pine | 0.17 | 0.18 |
| Beech | 0.17 | 0.16 |
| Linden | 0.13 | 0.15 |
| Spruce | 0.15 | 0.13 |
| Oak | 0.12 | 0.10 |

should note the humidity of the air. Absorption depends more strongly on humidity than refractive index, but for comparison of properties humidity is important.

**Additional tasks**

1. Estimate the thickness of the wood with a given birefringence and without absorption where the wood plate would function as a $\lambda$-quarter plate ($\delta = \pi/2$) or a $\lambda$-half plate, for the wavelength of your microwave emitter. Use your experimental data or data from table 4.2 [7].
2. If the incident beam is polarized at 45° with respect to the wood fibres of a wood plate with thickness as calculated in the previous task, the transmitted microwave is circularly polarized. This means that rotating the receiver's horn would not change the signal. For the thickness of wood from the previous task, calculate the elliptical state of the transmitted microwaves, if the absorption is included in the analysis. Use your experimental data or data from table 4.2 [7]. Are the transmitted microwaves still circularly polarized?
3. The circular polarizer in optics consists of a linear polarizer glued onto the $\lambda$-quarter plate. As microwaves are always polarized, the $\lambda$-quarter plate satisfies. To obtain the circularly polarized microwaves, the angle between the incident polarization direction and the wood fibres have to meet certain requirements. What angle is required for the piece of wood to act like a circular polarizer if the absorption is not negligible?
4. Implement the above as a project. Cut a piece of wood and measure its properties: the birefringence and the absorption coefficients. Calculate the thickness of the $\lambda$-quarter and $\lambda$-half plates and cut them from the same log. Measure the properties of such a plate. Measure the properties of the same plate in the winter and in the summer. Humidity strongly affects the properties of wood.

## 4.4 Anisotropic materials in visible light

Initial ideas about newly synthesized structures of liquid crystal usually come from observations of textures. The 'texture' often comprises the colourful appearance of a liquid crystal in the cell between crossed polarizers. Some nice examples of textures have already been presented in figure 1.2. Students observed textures when they assembled a cell with a liquid crystal and observed it between crossed polarizers under the microscope.

In this section, we intend to illustrate two fundamental optical phenomena, colours of an anisotropic material between crossed polarizers, often observed in liquid crystals, and double refraction, typical for anisotropic materials in general. Instead of liquid crystals, simple easily accessible materials like transparencies or cellophane and gems from anisotropic material are used. Regular school equipment usually also includes anisotropic crystals that are perfect for the purpose.

Let us first provide a theoretical background for the two phenomena we shall illustrate in this section.

If an anisotropic material is placed between crossed polarizers, often colours will appear [8]. Why? As discussed before, the linearly polarized light transmitted through a thin sheet of anisotropic material becomes elliptically polarized. The first parameter that defines the polarization state of transmitted light is the angle between the transmission direction of the polarizer and one of the principal axes of the anisotropic material. This angle defines the amplitudes of the electric field in both beams that the light splits in an anisotropic material. The second parameter that defines the polarization state is the phase difference $\delta$. The phase difference depends on the difference between the optical paths for the two perpendicularly polarized beams. In previous sections, the dependence on birefringence and thickness for liquid crystals and for wood were discussed for monochromatic microwaves and light sources. If the sample is illuminated by white light, each wavelength has a different optical path through the sample because

$$\delta = 2\pi\frac{\Delta nd}{\lambda}. \tag{4.22}$$

Here, as before, $\Delta n$ is the birefringence or the difference between refractive indices for a specific direction of light and for a specific wavelength. For simple materials, dispersion is not so important, therefore the dependence of refraction coefficients on the wavelength of light is considered negligible. The thickness of the material is marked by $d$ and the wavelengths marked, as usual, by $\lambda$. As a consequence, different wavelengths have different phase differences and they have different elliptical polarization states, as seen in the illustration in figure 4.20. The second polarizer absorbs light of different wavelengths differently. For crossed polarizers all wavelengths that have a phase difference of a multiple of $2\pi$ are completely absorbed. All partial beams with a phase difference of close to $2\pi$ are significantly absorbed. Other wavelengths are absorbed more or less, for example, if the incident light is polarized at 45° with respect to the principal direction of the polarized light, the partial beams with wavelengths close to odd multiples of $\pi$ are hardly absorbed at all. So, some wavelengths, that is some colour components in initially white light, are almost completely missing, some are more, and some are less absorbed, and the spectrum of light is significantly changed. Changing the spectrum of white light by wavelength dependent absorption can mean one thing only—the light becomes coloured. Which colour actually appears depends on several facts, the

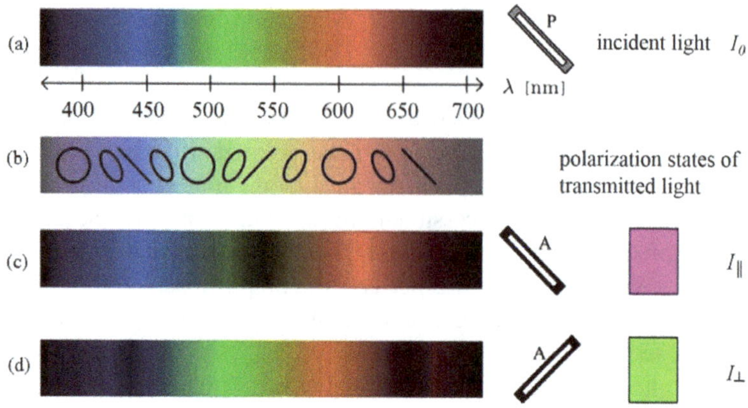


**Figure 4.20.** (*a*) The spectrum of the incident light is presented by corresponding colour components. P marks the polarization of the incident light. (*b*) The polarization state of transmitted light depends on the wavelength. States for different wavelengths are illustrated by characteristic ellipses. (*c*) and (*d*) Coloured illustrations of transmitted light spectra after analysers oriented as marked by symbol A. Corresponding colours are given on the right. After [8]. Note that wavelengths absorbed for crossed polarizers are not absorbed (bright) for parallel polarizers and vice versa.

mutual orientation of the polarizer and analyser, the thickness of the sample and the birefringence. However, when a material is chosen, the birefringence for an orthogonally incident light is fixed but still, the phase difference is different for different wavelengths resulting in different elliptical states.

The colours presented in the Michel–Levy chart (figure 2.23) appear due to the reasons discussed above. However, the transmitted light is not always coloured. If the sample is very thin, the phase difference is much smaller than $\pi$ and the condition for absorption is not fulfilled for any of the wavelengths. The sample is bright for parallel polarizers and dark grey for crossed polarizers. For thicker samples however, the condition for absorption for either parallel or perpendicular polarizers can be fulfilled for more than one wavelength, and the colours appear. For very thick samples, the light for several wavelengths is absorbed and as a result, the colour does not change but the intensity of the transmitted light is smaller.

Students will study all these effects in simple ways. As anisotropic materials, various foils or tapes are used. The materials are easily accessible and if the lecturer has several polarizers available, the lecture can be presented in an active way, with students also doing simple experiments within a large audience.

Another phenomenon that is characteristic for anisotropic materials is double refraction. It is also shown for liquid crystals by using a wedge cell but the phenomenon is much easier observed using anisotropic crystals often available as gems in various shops selling ‘positive energy’, ‘astral waves’ and similar nonsense.

### 4.4.1 Colours of anisotropic materials in polarized light

The dependence of the phase difference $\delta$ is the most easily observed through spectra of transmitted light. An anisotropic material placed between crossed polarizers and illuminated by a white light is often coloured. Very thin samples are grey, samples a little thicker are yellow, but other colours could also be observed, even thicker samples are green or magenta, and finally, very thick samples are grey again. How the phase difference is determined by the thickness of the sample, the influences of birefringence and the wavelength, equation (4.22), and the colour of the transmitted light is rather easy to study. To limit the study to the dependence of the phase difference on the wavelength, only the sample with a constant thickness and a constant difference between the refractive indices is used. Illuminating the sample with a white light with a continuous spectrum, sunlight for example, the spectrum of the transmitted light, in comparison with the spectrum of the incident light, clearly shows which wavelengths are absorbed more and which less. For crossed polarizers, the wavelengths with wave differences close to an even factor of $\pi$ are the ones that are absorbed more, and for parallel polarizers the same is true for wavelengths with the phase difference close to an odd factor of $\pi$. Comparing the spectra, one clearly sees that both spectra are exactly complementary. Furthermore, comparing the spectra, the relation between spectra and colours becomes straightforward.

**Experimental aims**

- To observe colours of plastic foils (transparencies, cellophanes, tapes) between polarizers.
- To observe, through a sample between polarizers, the dependence of the spectrum of transmitted white light on orientation.
- To observe the dependence of the colours of the plastic foils on thickness.
- To observe the spectrum of transmitted light through crossed and parallel polarizers, and to compare the differences for different thicknesses.
- To calculate the birefringence of the material.
- To analyse possible principal directions in a general plastic sample, for example, a transparent plastic cover, a CD box, etc.

**What is required for the experiments**

- Pieces of plastic foils, celophanes or tapes.
- Two polarizers.
- A spectrometer.
- Various transparent plastic items.
- A tablet or a portable computer for which it is possible for the screen to be horizontal.

Take two polarizers and put one above another. Orient one with respect to the other in such a way that the light is absorbed, and they are crossed. Insert between the polarizers various thin transparent anisotropic objects. Observe them. Slowly rotate the samples between crossed polarizers. If they are coloured, observe how the colours change. Find the two orientations of a sample where the sample is dark. Mark these two orientations on the sample. Now, the orientation of the sample can be measured with respect to those two axes. For which orientation are the colours the most intense, and the sample the most colourful? Note down the colour.

Next, rotate one of the polarizers by 90° so that the polarizers are now parallel, and rotate the sample again. What does the sample look like if it is oriented in one of the marked orientations? What does it look like if it is oriented in the 'most colourful' orientation? Note down the colour and compare it to the most intense colour of the sample between crossed polarizers.

Choose one of the samples that is yellow or blue between crossed polarizers. First measure the spectrum of incident light. Compare the spectrum of transmitted light with the spectrum of incident light. Measure the wavelengths where the transmitted light is greatly absorbed. Rotate one of the polarizers by 90° so that the polarizers become parallel and the orientation of the sample is maintained. Compare the spectrum of the transmitted light through the sample between parallel polarizers oriented in the spectrum of the transmitted light through perpendicular polarizers. Determine the wavelengths where the light is greatly absorbed.

Take the second thin anisotropic foil and determine the orientations for which the foil is dark between crossed polarizers. Put the second layer of the same thin foil onto the first one, oriented the same way. Observe the colour between crossed and parallel polarizers, measure their spectra and calculate the birefringence of the sample for a known foil thickness and known wavelengths of the minima for the transmitted light through the foils for crossed and parallel polarizers. A few examples of such measurements are presented in figure 4.21.

Finally, use a white computer screen as a source of polarized light. Lay on a screen a transparent object and observe it through the polarizer. If the object is made from an anisotropic material, colours usually appear. From the orientation of the polarizer for which the light from the screen is absorbed, principal directions at different places can be determined. Rotating the sample on the screen and marking the principal orientation at different positions, one can get an idea about the 'optical' structure of the sample.

Components of white linearly polarized light transmitted through an anisotropic material are elliptically polarized to various extents, determined by two factors: the ratio between the components of the electric field along the principal directions in an anisotropic material and the wavelength of light. For the same sample, if it is thick enough, several wavelengths can be completely absorbed due to the phase differences that are multiples of $2\pi$. It is easily shown that dark areas are always an indication of the lack of anisotropy due to a special direction of order, or due to a lack of it, or due to the correspondence of principal directions to the transmission directions of the polarizers. Other areas are coloured. For any other mutual orientations of polarizers, the light is never completely absorbed and the determination of several optical properties hinting at the underlying structure is not possible.

This is nicely demonstrated by a plastic cover on the screen, see figure 4.22. The cover is made from a polymeric material that is squeezed into the shape of a cover. During squeezing, polymers orient and the material becomes anisotropic with the principal direction changing from one place to another.

*Tricks of the trade*. For all experiments, the best choice is daylight or an incandescent light source. LED and other fluorescent light bulbs have several discrete components in the spectrum, therefore the comparison of the incident and transmitted spectra is not straightforward. An LCD as a light source, however, has three discrete

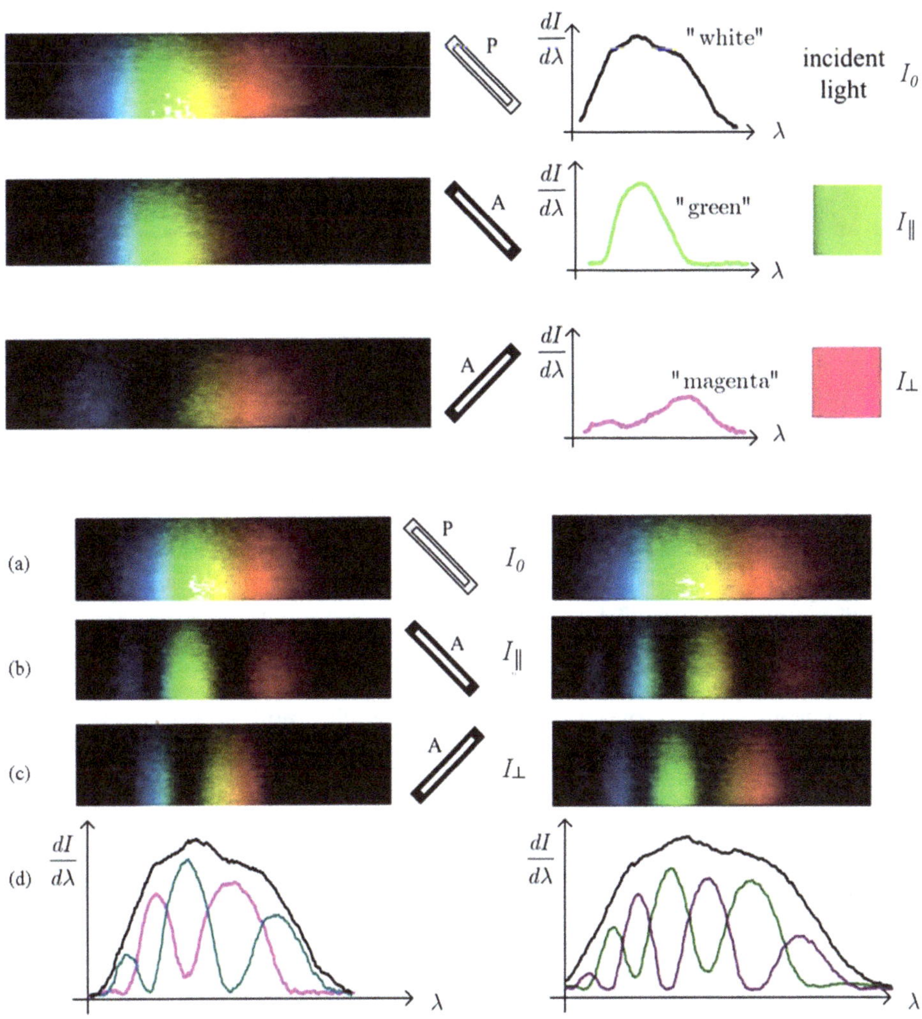


**Figure 4.21.** Rainbow-like pictures are taken on the screen after the diffraction grating. The white light spectrum contains all colours of the rainbow, while the light passing through the polarizer (transmission direction indicated by P), the anisotropic materials and the analyser (transmission direction indicated by A) have dark lines for some of the wavelengths. Graphical presentations of the same spectra are given. Top: The light was transmitted through the two layers of transparent tape of thickness $d$ = 55.2 μm. Below left: Through four layers of tape. Below right: Through six layers of tape. Note that in thicker samples, the condition for absorption is fulfilled for different wavelengths at the same time. Note also, the transmitted light spectra for crossed and parallel polarizers are complements.

components only. Therefore a screen as a light source for measurements of transmitted spectra is not appropriate, but for observations of textures it serves perfectly.

For experiments with controlled thicknesses, like when tapes or foils are placed one on top of another to obtain variations in thickness, use a material of a thickness that appears yellow between crossed polarizers. This means the shortest blue wavelengths are more absorbed than the rest of the wavelengths, but the condition for a full

(a)

(b)

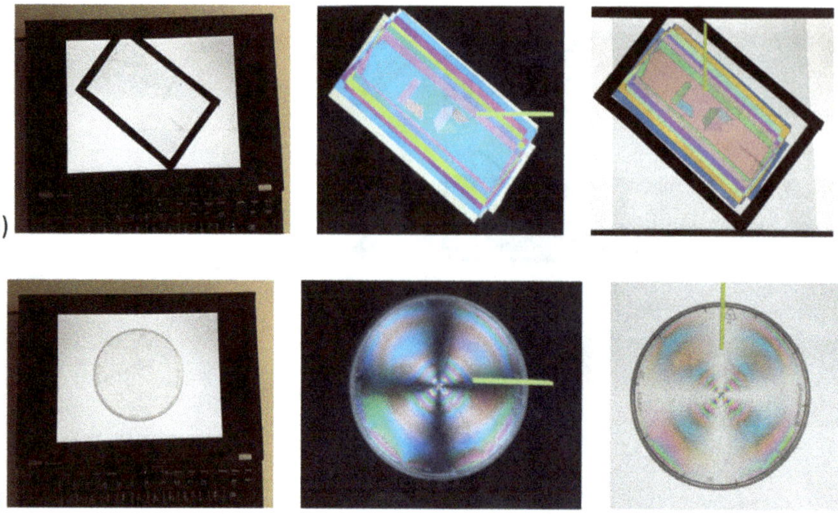

**Figure 4.22.** (*a*) From left to right: various layers of tapes on the screen; observed through a polarizer perpendicular to the polarization of light emitted from screen; observed through a polarizer parallel to the polarization of light emitted from screen. (*b*) From left to right: a plastic cover on the screen; observed through a perpendicular polarizer; observed through a parallel polarizer. The green line indicates the transmission direction of the polarizer placed on the tapes or on the plastic cover.

absorption is not fulfilled for any of wavelengths. Such conditions are met with more layers and the concept of thickness influencing colour is easier to comprehend.

*Suggestions for the lecturer.* When observing liquid crystals in a cell, all the phenomena studied in this set of experiments is combined. Some places are dark, and this is possibly simply due to the absence of a liquid crystal in the cell (usually seen as round dark spots) or liquid crystals are oriented in a way that the principal direction corresponds to the transmission direction of one of the crossed polarizers. These points hint at the orientation of long molecular axes. However, different colours may mean different states of order or different orientations of long molecular axes. By rotating the sample between crossed polarizers, several hints at the structure can be obtained. A nice example is the study of a plastic cover between crossed polarizers. Everything studied on the plastic cover can be applied later to the study of a liquid crystal.

**Additional tasks**

1. Measure the spectra of various thicknesses of the same type of a transparent foil; determine the wavelengths of extremal absorption and calculate the birefringence. Make a table of birefringence for various transparent anisotropic materials and compare them.
2. Measure the spectra of transmitted light upon rotation of one of the polarizers and upon rotation of the sample. Analyse the results on the basis of what is known from the section about elliptical states of light.
3. Construct a 'polarized light picture'. Some artists make polarized light pictures that hide two images—one for one direction of a polarizer and the other one for a polarizer perpendicular to it. An example can be seen in figure 4.23. Try and work out the rules for how such a picture is constructed.

**Figure 4.23.** The artistic polarized light picture (*a*) without polarizers, (*b*) through a polarizer with vertical transmission direction and (*c*) through a polarizer with horizontal transmission direction.

### 4.4.2 Double refraction in crystals

If a material is transparent and anisotropic, light propagates in each direction with two different wave velocities associated with two mutually perpendicular polarizations. For obliquely incident light, the light refracts differently for each polarization. The phenomenon is called double refraction. Double refraction is nicely demonstrated by the propagation of a narrow laser beam through a prismatically shaped anisotropic material, as the splitting of a beam is evident. However, one can also observe other consequences of double refraction; for thick enough samples, images of objects observed through an anisotropic material can appear duplicated.

**Experimental aims**

- To observe a duplicated image through an anisotropic crystal.
- To determine the polarization direction of light for each of the two images.
- To observe the crystals illuminated by a polarized light through a polarizer.
- To determine the principle directions of the crystal.

**What is required for the experiments**

- Anisotropic crystals, usually available as gems.
- A polarizer.
- A horizontal portable computer screen or a tablet.

The crystal is placed over some text and the text is observed through it. The text is observed perpendicular and at an angle. The crystal is rotated and the positions of both images are observed.

Next, the polarizer is placed above the crystal and the text is observed through the polarizer and the crystal, see figure 4.24. The polarizer is rotated and for one orientation of the polarizer one of the images disappears. If the polarizer is rotated a further 90° the second image disappears and the first reappears. With this experiment, students verify that the initially non-polarized light reflected from the paper splits into two beams that enter the eyes from slightly different directions.

It is a different story when an incident light is already polarized. Easily available sources of polarized light are liquid crystalline screens of portable computers or tablets. When an anisotropic crystalline gem is placed on the text on the screen, again a double image occurs. However, rotation of the crystal results in the disappearance of one or the other image, even without a polarizer. Adding the polarizer, oriented in a way that the initially bright computer screen is not seen

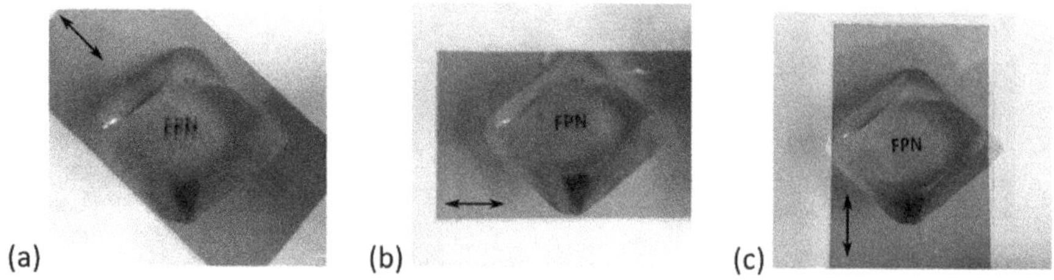


**Figure 4.24.** (*a*) The anisotropic crystal over the text. The polarizer is at a general angle with respect to the principal polarizations. (*b*) The view through a polarizer with the transmission direction parallel to one principal polarization. (*c*) The view through a polarizer with the transmission direction parallel to the other principal polarization (Photo: Maja Pečar).

through the polarizer i.e., the light from the screen is polarized perpendicularly to the transmission direction of the polarizer, the crystal only appears bright at orientations where duplicated letters are seen. In orientations where only one image is observed, the whole crystal is also dark and letters cannot be seen.

The above experiments clearly show that the light relating to both images is polarized, and furthermore, the polarization in the two beams is perpendicular to each another. When the transmission direction of the polarizer coincides with one of the polarization planes, the other beam is completely absorbed and only one image is seen. However, if the incident light is polarized and the polarization direction has a general angle with respect to principal directions in the crystal, the light again splits into two beams, but now both beams have different amplitudes of electric field. The amplitude is defined as a projection of the electric field to the direction of the principal axes. This actually means that the intensity of the transmitted light for each beam changes with rotation of the crystal.

*Tricks of the trade.* The effects of dimming one or other of the duplicate images is best seen if the letters that represent the observed objects are white on a black background. Black letters on a white background do not show the change in intensity as they are already black, but the light from the white background is rather intense.

*Suggestions for the lecturer.* All the experiments with anisotropic crystals can be carried out in a proactive way. Students experiment with polarizers, crystals and computer screens, and with the help of the lecturers' questions can discover the rules for double refraction. Before starting the experiment, the polarization of the electromagnetic wave should be refreshed.

**Additional tasks**

1. Find the principal directions for the gem. For the planes in which principal directions are found, the gem seems dark between crossed polarizers.
2. If the gem material is known, find the data for refractive indices. Measure the dimensions of the crystal and calculate the shift of both beams for different angles of incidence. Do experimental and theoretical results based on the refraction of wave vectors coincide? Read a discussion on energy propagation in anisotropic materials in an optics textbook.

3. Combine transparencies or celophanes and an anisotropic gem. Observe text on paper or on a screen through combinations: the crystal on the transparency, and the transparency on the crystal etc.

## Bibliography


[1] Ziherl S, Bajc J, Urankar B and Čepič M 2010 Anisotropy of wood in the microwave region *Eur. J. Phys.* **31**(3) 531–42

[2] Ziherl S, Bajc J and Čepič M 2013 Refraction and absorption of microwaves in wood *Eur. J. Phys.* **34**(2) 449–59

[3] Perkalskis B S and Freeman R J 1999 Tensors in the lab—The thermal resistivity of wood *Am. J. Phys.* **67**(5) 452–5

[4] Čepič M 2012 Knitted patterns as a model for anisotropy *Phys. Educ.* **47**(4) 456–61

[5] de Gennes P G and Prost J 1995 *The Physics of Liquid Crystals* International Series of Monographs on Physics (Oxford: Clarendon Press)

[6] Ciferno T M, Ondris-Crawford R J and Crawford G P 1995 Inexpensive electrooptic experiments on liquid crystal displays *The Physics Teacher* **33**(2) 104

[7] Ziherl S 2013 *Anisotropic properties of wood in microwave region* PhD thesis, Faculty for Science and Mathematics, University of Maribor, Slovenia, 5

[8] Babič V and Čepič M 2009 Complementary colours for a physicist *Eur. J. Phys.* **30**(4) 793–806

# Chapter 5

# Elementary teaching module

The experiments suggested in this book aim to provide experiences for students coming across liquid crystals and their physics in greater detail for the first time. One could reserve a whole semester of lectures and laboratories for the topic of 'liquid crystals'. However, the topic is interesting and important enough to be introduced to a general audience. An expert in liquid crystals would have in mind a large selection of experiments with liquid crystals to support the understanding of typical phenomena. However, if a lecturer who is an expert in other fields, has an interest in introducing the subject of liquid crystals to a general audience, here is a brief description of a short teaching module that was developed and tested at the Faculty of Education, University of Ljubljana.

## 5.1 What should students learn about liquid crystals?

The aim of the teaching module is to explain how a liquid crystal display functions and the a role of liquid crystals in its operation. A basic understanding of LCDs is general knowledge as displays are an important example of a link between academic research and applications. If non-physics students come across such an example at least once during their studies, they may not consider academic research to have no immediate specific application or to be obsolete.

What should students know about liquid crystals and what might be interesting to them? In my opinion, the following specific concepts deserve consideration:

- To understand the function of a display, students should be able to recognize and identify the object of interest, the pixel, on an enlarged screen.
- Students should be aware that liquid crystals are a special phase of matter with very special properties.
- Students should become familiar with the following concepts that are important for transparent anisotropic materials: anisotropy, double refraction and birefringence.

doi:10.1088/978-1-6270-5300-6ch5

- Students should be aware that liquid crystals must be ordered to exploit their special properties.
- Students should know that liquid crystals are easily manipulated by external stimuli, such as surfaces and an electric field.
- Students should link the concepts mentioned above in a consistent picture of pixel operation.

Students could consider in more details any of the topics presented earlier, if there is more time available. A combination of experiments with wood and experiments with liquid crystals would be very useful, if the schedule allows.

After setting the goals, consideration should be given to the knowledge that students require before starting the module, the topic has to be placed in the curriculum as it is usually not mentioned, and the methods for successfully cultivating this new knowledge should be chosen.

## 5.2 Basic concepts and placement in the curriculum

As liquid crystals are materials that form a special liquid crystalline phase, students should be familiar with the concept of phases and they should know that phase transitions occur at an exact temperature. The next important concept is the speed of light in a transparent medium, its relation to the index of refraction and Snell's law. Students should also know that materials become electrically polarized in an external electric field and that a material's polarization is a consequence of structural changes of material in the electric field.

Consideration of the required preliminary knowledge gives some hints about the placement of the subject. Liquid crystals can provide a motivational pathway through several topics in physics.

When teaching thermodynamics, or more specifically, phases and phase transitions, an additional phase is shown in liquid crystals. The appearance of liquid crystals in their liquid crystalline phase is significantly different from their solid or liquid states. This is not the case for other materials; for example, a ferromagnetic material or a superconductor looks exactly the same when the phase transition to the ferromagnetic or superconductive phase appears at the transition temperature. The phase transition has to be deduced from other properties and it is not so educational.

When Snell's law is introduced and prisms are discussed, a rainbow is often added as an interesting phenomenon that is observed because the speed of light depends on its wavelength. A similar phenomenon, i.e. the dependence of the speed of light on the polarization of light, can be discussed, and it could be illustrated by double refraction, as described in subsection 4.4.2. A demonstration experiment can be carried out during lectures and detailed measurements of refractive indices taken during the laboratory. Explanations of why it is so can be left to the lectures on liquid crystals and during optics they serve for motivation only.

A picture element (pixel) is formed by confining nematic liquid crystal between two conducting plates. One pixel is thus a capacitor with a dielectric material (liquid crystal) between the plates. When voltage is applied to the cell surfaces (capacitor plates), the material between the plates polarizes. Electric polarization leads to changes in the structural properties of the materials, that is, the reorientation of molecules, affecting the transmission of light through the cell. Thus, the function of the pixel can be used to consider electric polarization in other materials.

## 5.3 Teaching module

The teaching module provides a member of the public, with at least a slight interest in science and technology, with the basic knowledge about liquid crystals necessary for understanding liquid crystals and LCD technology. The choice of methods for the teaching intervention is based on the fact that liquid crystals are usually a very a new topic for students and they have to be introduced from the very beginning. Several concepts are introduced to students for the first time and students are required to build up their understanding without any preliminary knowledge. Therefore this module provides a combination of a traditional lecture along with several demonstrations providing personal experiences to students, where most of the fundamental concepts and properties are introduced: a chemistry laboratory, where students synthesize a liquid crystal; and a physics laboratory, where they use the product they have synthesized in the chemistry laboratory, to study various physical properties of liquid crystals using an active learning approach. The laboratory work allows students to construct and to comprehend several new ideas that are all linked together in the application, a liquid crystalline display. The estimated time for each part is 90 min.

### 5.3.1 Lecture

The lecture provides fundamental information about liquid crystals, about their properties and how they are used in applications. After the lecture, students should be able to

- List some products based on liquid crystal technology.
- Identify the additional liquid crystalline phase and phase transition.
- Describe and illustrate the structure of liquid crystals on a microscopic level.
- Test the property of liquid crystals which is important for applications: birefringence, resulting from the orientational ordering of molecules and its effect on the transition of light.
- Test the above-mentioned properties of an unknown substance.
- Explain the effect of an electric field on molecular orientation.
- Explain how LCDs work.

If necessary, a short part of the lecture introduces polarizers and their properties. The lecture starts with a magnification of an LC screen as a motivation; it is explained that at the end of the module students will be able to understand how the display works. The lecture continues with a description and a demonstration of the new, liquid crystalline, phase, which is similar to an opaque liquid in its macroscopic appearance. All three phases (crystalline, liquid crystalline and isotropic liquid) are shown while heating a sample, as a demonstration. The microscopic structures of all three phases are presented by schematic diagrams and the concept of orientational order is introduced. The molecular shape allowing orientational ordering is discussed. The concept of light propagation in an anisotropic material is introduced and double refraction is shown using a wedge liquid crystalline cell. Polarizations of both beams are determined.

In LCDs, the electric properties of molecules are very important so the effect of the electric field on the molecular orientation is discussed as well. Electric field induces electric dipoles in molecules and consequently the molecules are rotated by an external electric field. Because the anisotropic properties depend on the structure of the liquid crystal in the cell, the transmission depends on the applied electric field. This leads to the structure of a pixel and how LCDs work. At the end of the lecture some interesting facts are mentioned, such as the fact that liquid crystals are also found in spider's threads and cell membranes in living organisms.

### 5.3.2 Chemistry lab

Laboratory work in chemistry is devoted to the synthesis of the liquid crystal MBBA. The aims of the laboratory work in chemistry are the following:

- Students learn how to synthesize the liquid crystal MBBA.
- Students realize that the product of the synthesis is useful for experiments showing the basic properties of liquid crystals.

Students synthesize the liquid crystal MBBA in a school laboratory from 4′-n-butylaniline and 4-methoxybenzaldehyde [1, 2] as described in subsection 2.1.1. Due to safety reasons, the synthesis has to be carried out under a fume hood.

### 5.3.3 Physics lab

Four experiments are performed during the physics laboratory work. They provide students with personal experiences and allow them to investigate the most important liquid crystalline properties.

*Experiment 1: An additional phase and phase transition*

**Experimental aims**

- Students learn that the liquid crystalline state is one of the states of matter.
- Students are able to describe the difference between the melting temperature and the clearing temperature.
- Students are able to measure the two temperatures and use them as a measure of success for the synthesis; if both temperatures are close to the temperatures given in the published data, the synthesis is successful.

Students use a water bath to heat the test tube with a frozen liquid crystal MBBA. They measure the temperature of the water assuming that the small sample of the liquid crystal has the same temperature as the bath. They observe how the appearance of the substance changes while heating the water bath. They measure the melting

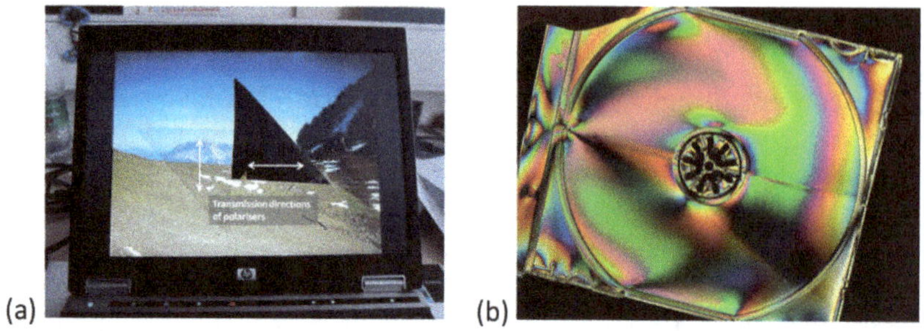

**Figure 5.1.** (*a*) The light emitted from an LCD can be used as a source of polarized light. (*b*) A transparent CD box is placed on the LCD providing a white background. If observed through a polarizer with transmission direction perpendicular to the polarization of the light emitted from the LCD, intense colours appear.

temperature at which the sample begins to melt and the clearing temperature at which the milky appearance of the sample starts to disappear.

*Experiment 2: Polarization*

This experiment is not necessary if students are already familiar with the concept of linearly polarized light and the function of polarizers.

**Experimental aims**

- Students learn what polarizers are and how they affect non-polarized light.
- Students learn how light propagates through the system of two polarizers.
- Students are able to test whether the light is polarized, and in which direction it is polarized, by using a polarizer with a known tranmission direction.
- Students are able to test whether the substance is optically anisotropic by using two polarizers.

From this activity students learn how to use a polarizer as an analyser. They also verify that the reflected light is partially polarized. Students observe various transparent materials placed between crossed polarizers and find that light cannot be transmitted when isotropic materials like water or glass are placed between the crossed polarizers. When anisotropic material such as scotch tape, cellophane or a CD box is placed between the crossed polarizers, light is transmitted. Colours are also often observed (figure 5.1). The activity provides experience that is used later in observations of liquid crystals in a cell.

*Experiment 3: Double refraction*

**Experimental aims**

- Students learn that birefringence is an important property of matter in the liquid crystalline state.
- Students are able to make a planar wedge cell and find an area where liquid crystal is ordered enough that the laser beam splits into two separate beams. the beams are observed as two light spots on a remote screen.
- Students learn how to verify the polarization of both beams using a polarizer.

Students manufacture a wedge cell from a microscope slide, a cover glass, foil for food wrapping or tape, and the liquid crystal MBBA [3]. Special attention is given to the rubbing of the microscope and cover glasses which enables the anchoring of liquid crystal molecules. Students direct the light onto the wedge cell (they use a laser pointer as the light source) and find the area of the cell where the laser beam splits into two beams. By rotating the polarizer between the cell and the screen, they verify light polarization in the beams. Then students heat the wedge cell with a hairdryer and observe the collapse of the two bright spots into one at the phase transition to the isotropic liquid.

*Experiment 4: Colours*

**Experimental aims**

- Students learn how to fabricate a planar cell, i.e. a cell with parallel glass surfaces.
- Students learn that liquid crystals are optically anisotropic and that light is transmitted if a cell filled with liquid crystal in its liquid crystalline phase is placed between two crossed polarizers; they should know that under such circumstances colours may also appear when the sample is illuminated by white light.
- Students learn that colours observed under perpendicular and parallel polarizers are complementary.
- Students learn how to mechanically order molecules in a planar cell.

Students manufacture a planar cell filled with the liquid crystal MBBA from a microscope slide, a cover glass and foil for food wrapping or tape as a spacer [3]. They observe the planar cell under a polarizing microscope (a school microscope with $M = 40$ or 100 and two crossed polarizing foils). They rotate the polarizing foils and observe how colours change. This experiment also illustrates the concept of complementary colours [4]. Afterwards, students heat the cell and observe colour changes. Next, an experiment is performed in which cells with ordered molecules are fabricated. Students make micro notches on a microscope slide by rubbing it with velvet soaked in alcohol. Molecules orient with their long axes parallel to the surface and the rubbing direction. A similar process is used in the fabrication of LCDs. They observe the cell with the ordered liquid crystal under the polarizing microscope. Finally, the cell is heated by a hairdryer and students observe the phase transition that appears as a dark front moving through the sample ending in a dark image when the liquid crystal between crossed polarizers is in the isotropic phase. The laboratory work concludes with a discussion of how a pixel in an LCD works, relating the changes of the liquid crystalline structure due to the electric field to the transmission rate of the pixel. At this point, the fact that colour filters are responsible for the colours of each part of the pixel is emphasized.

## 5.4 Evaluation

Although the results of the evaluation of the module presented above are not particularly in line with the sense of this textbook, the evaluation of the newly introduced topics will give important feedback to the lecturer. Therefore, an evaluation of the results of the pilot study of the introduction module is briefly presented here.

The aim of study was the evaluation of a teaching module for non-physics students, which could also be implemented in primary schools students. The idea was to give future high school teachers, who are non-specialists in science subjects, a basic knowledge of liquid crystals, so they are able to answer the potential questions of younger students when they are teachers themselves. The teaching module was preliminarily tested by a group of 90 first-year students enrolled in a four-year university programme for primary school teachers at the Faculty of Education (University of Ljubljana, Slovenia) in the school year 2010/11. The evaluation of the module considered the efficiency of the teaching intervention, which concepts students had assimilated and comprehended and to what extent.

Participants of the study were first-year pre-service teachers (future primary school teachers). They were chosen because their previous knowledge about liquid crystals is just as negligible as that of students from other faculties and study programmes [5]. In addition, pre-service teachers do not have any special interest in natural sciences, but they have to be as scientifically literate as everyone else who has finished high school. And most importantly, the pre-service primary school teachers form the only homogeneous group that has physics included in the study programme and was, in the circumstances, at least approximately, large enough to allow for a quantitative study. In the group of 90, six were male and 84 were female students. They were, on average, 20.1 years old and they achieved 19.7 points out of 34 in the final examination at the end of the upper secondary school. The average achievement in the final examination in Slovenia was 19.5 points out of 34 and a total of 8842 candidates attended the final examinations in Spring 2010. The group studied consisted of a predominantly rural population with mixed socio-economic status. This qualifies the participants as average representatives of the population just after finishing upper secondary school programs.

The data was collected in the following way: through a pre-test; classroom observations of the group work; worksheets; and tests, in spring 2011. The pre-test had 28 short questions. The first part (seven questions) related to general data about the student: gender, age, secondary school, final examination results, residence stratum and motivation towards science subjects. The second part (19 questions) related to liquid crystals, their existence, properties and microscopic structure. The pre-test was applied at the beginning of the lecture relating to liquid crystals. Those students who did not attend the lecture filled in the pre-test before the beginning of the compulsory laboratory work in chemistry. The worksheet for the laboratory work in chemistry included a procedure to synthesize liquid crystal MBBA, a reaction scheme, observations and conclusions regarding the synthesis and questions from chemistry relating to liquid crystals and the laboratory work. The worksheet for the laboratory work in physics presented the properties of polarizing foils and optically anisotropic materials and experiments with the liquid crystal MBBA. Test 1 included 17 short questions relating to the knowledge obtained during the lecture and laboratory work. It was taken immediately after the end of the physics laboratory (in May 2011). Test 2 was part of an examination held four weeks later (June 2011). It had 17 questions that, again, covered the contents of the lecture and the laboratory work. The questions on test 2 were similar to those on the pre-test and test 1.

Let us give a short account of the results. 150 students attended compulsory lectures and laboratories. They worked on the synthesis a week after the lecture, and another week later they carried out the experiments with liquid crystals at the physics laboratory. Students worked in groups of three or four in the chemistry laboratory and in pairs in the physics laboratory. However, all data (tests and worksheets) were collected for only 90 students; and only their achievements are presented. All groups made the synthesis successfully, which was confirmed by measuring the melting and clearing temperatures of the synthesized liquid crystal MBBA. On average, 63.3% of worksheets were correctly filled in. All the experiments were successfully carried out in the physics laboratory. The only difference was that students did not prepare the wedge cell by themselves. Due to the lack of time, cells were prepared in advance. On average, 84% of worksheets included correct answers to questions and observations. In the pre-test students, on average, achieved 24.0% of the available points. Their achievements show that their prior knowledge about liquid crystals was limited, as expected. In test 1, which was held immediately after the physics laboratory, students achieved 68.1% on average. Test 2 was part of a regular examination in physics. In test 2, students on average achieved only 63.5% of the available points.

Figure 5.2 shows the distribution of students versus the points achieved on the pre-test, test 1 and test 2. The percentage of students who achieved higher scores in tests 1 and 2 than in the pre-test is evident. The expected level of knowledge about

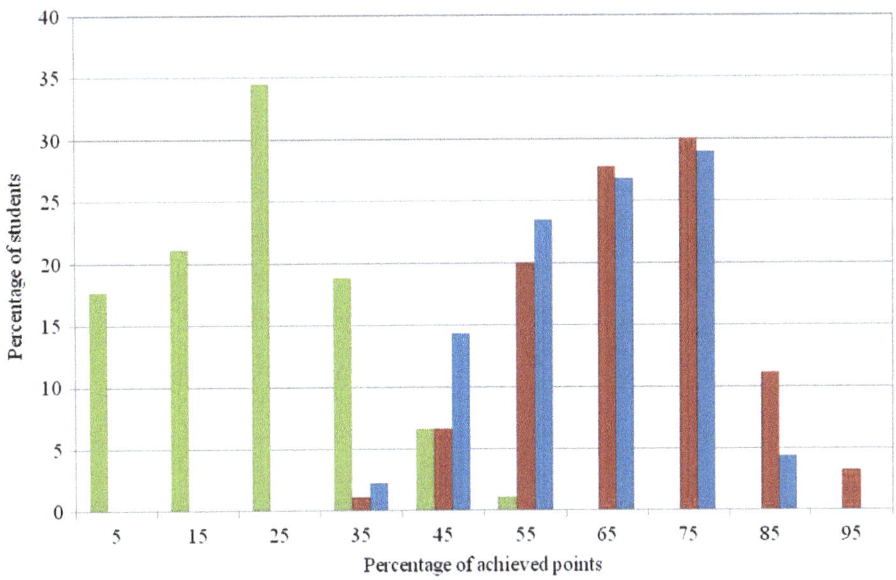


**Figure 5.2.** Distribution of the percentage of students versus the percentage of achieved points in tests. Green: pre-test; red: test 1; blue: test 2.

liquid crystals is highest immediately after the activities. From figure 5.2, it can be seen that most students achieved less than 50% of available points in the pre-test. In test 1, most of the students achieved over 50%. It can clearly be seen that the percentage of students achieving a higher percentage of points in test 2 is lower than in test 1, when the impressions of the laboratory work were still fresh in the mind.

Results show that even in modest circumstances, with limited time available, it is possible to introduce a new topic to students. It is enough for them to grasp basic ideas and underlying concepts. However, for a deeper comprehension more time is required, as well as greater effort from the students.

## Bibliography

[1] Liberko C A and Shearer J 2000 Preparation of a surface-oriented liquid crystal: An experiment for the undergraduate organic chemistry laboratory *J. Chem. Educ.* **77**(9) 1204–5

[2] Patch G and Hope G A 1985 Preparation and properties of cholesteric liquid crystals *J. Chem. Educ.* **62**(5) 454–5

[3] Pavlin J, Vaupotič N and Čepič M 2013 Liquid crystals: a new topic in physics for undergraduates *Eur. J. Phys.* **34** 1–17

[4] Babič V and Čepič M 2009 Complementary colours for a physicist *Eur. J. Phys.* **30**(4) 793–806

[5] Pavlin J, Vaupotič N, Glažar S A, Čepič M and Devetak I 2011 Slovenian pre-service teachers' conceptions about liquid crystals *Eurasia* **7**(3) 173–80

# Chapter 6

## Further reading

The textbook started with a brief description of liquid crystals and several experiments illustrating their properties, concepts relating to liquid crystals, and explanations of applications that use liquid crystals. Any of the activities may stimulate the curiosity of students but lecturers might want more details about specific questions or they may want to include more experiments in laboratory. This concluding chapter briefly looks at the literature on the subject. We start with books; there are many but we will consider only a few of them. Next, there is a short report on the university level literature about liquid crystals. Most of the discussion focuses on papers that have been published in *European Journal of Physics* and *American Journal of Physics*. We will not focus here on papers that have already been extensively cited in previous chapters, we aim to suggest papers that offer new and interesting views or suggest more advanced experiments. Finally, we briefly report on liquid crystals and their use in high school, and at lower levels of education. These mainly appeared in *Physics Education* and *The Physics Teacher*.

A brilliant introduction to liquid crystals is the textbook by Peter Collings, Liquid Crystals: Nature's Delicate State of Matter [1]. The book covers all general aspects relating to liquid crystals and their applications. It is written for a novice in the field, explaining the concepts by reasoning and discussion.

For readers who desire a thorough study, the textbook by de Gennes and Prost, The Physics of Liquid Crystals [2] gives a thorough, detailed, theoretically and experimentally supported explanation of liquid crystals, various phases and their properties. Oswald and Pieranski have similar aims in their two books, The Liquid Crystals series: Nematic and Cholesteric Liquid Crystals [3] and Smectic and Columnar Liquid Crystals [4]. In these textbooks, the reader can find descriptions, explanations and theoretical considerations of various phases with studies of more complex structures than nematics. In addition, several concepts are accompanied by relatively simple experiments that can easily serve as an upgrade to simpler experiments presented here, in an advanced laboratory. If even these books are not

doi:10.1088/978-1-6270-5300-6ch6

enough, answers to almost any question can be found in the Handbook of Liquid Crystals [5]. Eight volumes cover all aspects relating to liquid crystals—materials, phase structures and their properties from experimental and theoretical points of view. The whole set is rather expensive and is not usually part of a faculty library. However, tables of contents are available on the web and authors of specific chapters are usually happy to send the latest working versions of their submissions to potential readers if they are requested.

For lecturers, another book is of special interest; Crystals That Flow: Classic Papers from the History of Liquid Crystals, prepared by Sluckin, Dunmur and Stegemeyer [6]. The book presents original papers form the time of discovery to the end of last century, many of them translated into English as they originally appeared in other languages. Papers are accompanied by comments, biographies of authors and other stories about the papers. The book gives an historical insight into the development of ideas and the understanding of concepts related to liquid crystals. An historical point of view is especially valuable for lecturers as they make interesting historical stories as *intermezzos* to sometimes difficult lectures, due to complexity of the topic.

Several papers regarding liquid crystals in education have also been published in journals. One can sort them into five classes: papers presenting the subject of liquid crystals to a broader audience, papers discussing theoretical descriptions of liquid crystals, papers discussing the chemistry and synthesis of liquid crystals, papers suggesting experiments for high school or introductory physics level, and papers suggesting more complicated and demanding experiments for advanced physics labs. When the reader is already familiar with liquid crystals to some extent, this is another set of articles that are interesting to read; we call them conceptual papers. They discuss more general physics concepts that liquid crystals could nicely illustrate. Let us briefly present some papers from those classes.

*Introductory papers* were published mainly in the 1980s and 1990s. Research into liquid crystals had already reached a high level by this time; several applications, such as simple displays were already on the market and the term 'liquid crystals' had become known to the public. The time seemed perfect to introduce the topic to physics and chemistry classes. However, papers were not organized as a lecture support but were mainly written as information for curious lecturers of chemistry and physics. Authors of [7–9] focused on liquid crystals from the perspective of their function in a display. All three general papers were meant for a high school teacher without preliminary knowledge of liquid crystals.

Several papers discuss liquid crystals theoretically. More papers of this type appeared in *European Journal of Physics*, but in *American Journal of Physics* extensive theoretical considerations were included in suggestions for advanced laboratory experiments. Theoretical considerations are worth considering when the lecturer needs theoretical illustrations for soft matter phenomena, for example: the behaviour of liquid crystals in external fields [10–12], fluctuations close to the phase transition [13], defects, their structure and their observation from a theoretical point of view [14–17], and the order parameters and their definitions [18]. This small set does not cover all theoretical papers, there are some that discuss mixtures, colloids etc as well; however, their focus is beyond the scope of this textbook.

As already mentioned, the synthesis of some liquid crystals is not very demanding. The chemistry of liquid crystals and their physical properties, from a chemist's point of view, are mostly discussed in *Journal of Chemical Education*. Papers mostly focus on simpler synthesis [19–22] and the presentation of various liquid crystals as a novel sort of material with new chemistry and peculiar properties [23, 24].

Most of the additional ideas for demonstrations, laboratories and projects may be found in papers that were aimed for high school and introductory physics courses. Very interesting ideas come from the time when research on liquid crystals started to bloom [25–27]. Experiments are mostly performed by rather simple means and are easily adaptable to experimental equipment used today [28]. Several elementary experiments are suggested that describe the properties of liquid crystals, some of them consider similar experiments, as discussed in this textbook, but with equipment that was available at that time. Materials exhibiting other phases, such as cholesterics, are often used, which lead to different results with known procedures and offer several opportunities for projects [29, 30]. Cholesterics are also used in several experiments discussing the effect of changing the Bragg reflection wavelength [31] by heating or cooling in foils containing a layer of a cholesteric liquid crystal. The colour of such foil changes upon the variation of temperature and foils serve as perfect detectors for the visualization of scalar temperature fields [32–36]. Several experiments allow students to study the function of a cell as a shutter in a display [37, 38] or in welders' glasses [39].

It seems that when researchers introduce some of their research results into lectures and laboratories, they are most motivated to write a professional article about their efforts when advanced experiments and the underlying theory is considered. Several ideas and suggestions for laboratory experiments can be found in *European Journal of Physics* and *American Journal of Physics*. Papers discuss light modulation with liquid crystals on an advanced level [40, 41], experimental evidencing of critical fluctuations close to phase transition [13] and other critical phenomena [42], magnetically induced birefringence [12], measurements of Freedericksz transition in an external electric field [11] and studies of the Kerr effect [43]. Some papers present experiments studying entirely new phenomena only recently discovered [44], like the giant non-linear response in liquid crystals doped with dyes.

Finally, to motivate students and to give them a wider perspective of the phenomena related to liquid crystals it is worth reading papers called *concept papers*. They discuss the place of liquid crystals among other new interesting materials with peculiar properties [45–47], the role of entropy in various systems [48–50], Landau theories and descriptions of phase transitions [51–54] and several other conceptual illustrations liquid crystals offer, or are useful for the understanding of liquid crystals.

I would like to mention my personal fascination with liquid crystals. I have been studying liquid crystals from a theoretical point of view for almost thirty years, and I have been trying to introduce them to teaching for more than 15 years. I have done a great deal of reading throughout my studies, but I still find new and interesting ideas for research, and for the transfer of this new knowledge to education. I suggest that everybody who finds liquid crystals interesting should read at least a few of the articles cited above, and they will discover new ideas in them.

## Bibliography


[1] Collings P J 2002 *Liquid Crystals: Nature's Delicate Phase of Matter* (Princeton, NJ: Princeton University Press)

[2] de Gennes P G and Prost J 1995 *The Physics of Liquid Crystals* International Series of Monographs on Physics (Oxford: Clarendon Press)

[3] Oswald P and Pieranski P 2005 *Nematic and Cholesterid Liquid Crystals: Concepts and Physical Properties Illustrated by Experiments* Liquid Crystals Book Series (London: Taylor & Francis)

[4] Oswald P and Pieranski P 2005 *Smectic and Columnar Liquid Crystals: Concepts and Physical Properties Illustrated by Experiments* Liquid Crystals Book Series (London: Taylor & Francis)

[5] Goodby J W, Collings P J, Kato T, Tschierske C, Gleeson H and Raynes P (ed) 2014 *Handbook of Liquid Crystals; 8 Volumes* 2nd edn (New York: Wiley-VCH)

[6] Sluckin T J, Dunmur D A and Stegemeyer H 2004 *Crystals That Flow: Classic Papers from the History of Liquid Crystals* Liquid Crystals Book Series (London: Taylor & Francis)

[7] Ball J E 1980 Liquid crystals *Phys. Educ.* **15** 108–9

[8] Hayn C H 1981 Liquid crystal displays *The Physics Teacher* **19**(4) 256–7

[9] Bradshaw M J 1983 Liquid crystal devices *Phys. Educ.* **18** 20–26

[10] Collings P J 1995 Liquid crystal displays *Am. J. Phys.* **63**(11) 1044

[11] Moses T 1998 The Freédericksz transition in liquid crystals: an undergraduate experiment for the advanced laboratory *Am. J. Phys.* **66**(1) 49–56

[12] Moses T, Durall B and Frankowiak G 2000 Magnetic birefringence in a liquid crystal: an experiment for the advanced undergraduate laboratory *Am. J. Phys.* **68**(3) 248

[13] Moses T, Reeves J and Pirondi P 2007 Fluctuations near a phase transition in liquid crystals *Am. J. Phys.* **75**(3) 220

[14] Fournier J B 1993 Nematic distortions around chiral and polar impurities: electrostatic and magnetostatic analogies *Eur. J. Phys.* **14** 184–8

[15] Fournier J B 1994 On the covariant elasticity of smectic-A phases *Eur. J. Phys.* **15** 319–24

[16] Repnik R, Mathelitsch L, Svetec M and Kralj S 2003 Physics of defects in nematic liquid *Eur. J. Phys.* **24** 481–92

[17] Kaučič B, Ambrožič M and Kralj S 2004 Interference textures of defects in a thin nematic film: an applet presentation *Eur. J. Phys.* **25**(4) 515–24

[18] Low R J 2002 Measuring order and biaxiality *Eur. J. Phys.* **23**(2) 111–17

[19] Liberko C A and Shearer J 2000 Preparation of a surface-oriented liquid crystal: An experiment for the undergraduate organic chemistry laboratory *J. Chem. Educ.* **77**(9) 1204–5

[20] Patch G and Hope G A 1985 Preparation and properties of cholesteric liquid crystals *J. Chem. Educ.* **62**(5) 454–5

[21] van Hecke G R, Karukstis K K, Li H, Hendargo H C, Cosand A J and Fox M M 2005 Synthesis and physical properties of liquid crystals: an interdisciplinary experiment *J. Chem. Educ.* **82**(9) 1349–54

[22] Jensen J, Grundy S C, Bretz S L and Hartley C S 2011 Sythesis and characterization of self-assembled liquid crystals: p-Alkoxybenzoic acids *J. Chem. Educ.* **88** 1133–6
[23] Brown G H 1983 Liquid crystals—the chameleon chemicals *J. Chem. Educ.* **60**(10) 900–5
[24] Ohline S M, Campbell M L, Turnbull M T and Kohler S J 2001 Differential scanning calorimetric study of Bilayer membrane phase transitions *J. Chem. Educ.* **78**(9) 1251–6
[25] Wright J J 1973 Optics experiments with nematic liquid crystals *Am. J. Phys.* **41**(2) 270–2
[26] Lalanne J R and Hare F 1976 Three liquid-crystal teaching experiments *J. Chem. Educ.* **53**(12) 793–5
[27] Nicastro A J 1983 Demonstrations of some optical properties of liquid crystals *The Physics Teacher* **21**(3) 181–2
[28] Ciferno T M, Ondris-Crawford R J and Crawford G P 1995 Inexpensive electrooptic experiments on liquid crystal displays *The Physics Teacher* **33**(2) 104
[29] Fergason J L 1970 Experiments with cholesteric liquid crystals *Am. J. Phys.* **38**(4) 425–8
[30] Lisensky G, Boatman E and College B 2005 Colors in liquid crystals *J. Chem. Educ.* **82**(9) 1360A–B
[31] Olah A and Doane J W 1977 Experiment in the Bragg reflection of light for the undergraduate using cholesteric liquid crystals *Am. J. Phys.* **45**(5) 485–8
[32] Edge R D 1993 String & sticky tape experiments: thermal conductivity and liquid crystal thermometers *The Physics Teacher* **31**(7) 412–13
[33] Bacon M E 1995 Heat, light, and videotapes: experiments in heat conduction using liquid crystal film *Am. J. Phys.* **63**(4) 359
[34] Pursell E E 2000 Infrared radiation: Herschel revisited *The Physics Teacher* **38**(9) 559–60
[35] Bochniček Z 2014 The visualization of infrared radiation using thermal sensitive foils *Phys. Educ.* **48**(5) 607–15
[36] Bochniček Z and Konečny P 2013 Thermal sensitive foils in physics experiments *Phys. Educ.* **49**(4) 419–24
[37] Crane H R 1983 Liquid crystal displays: watches, calculators and (soon) cars *The Physics Teacher* **21**(7) 467
[38] Becchetti F D and Murphy M J 1985 Projected liquid-crystal computer display *The Physics Teacher* **23**(6) 382
[39] Kraftmakher Y 2012 Demonstrations with a liquid crystal shutter *Phys. Educ.* **47**(3) 329–33
[40] Martínez J L, Moreno I and Ahouzi E 2006 Diffraction and signal processing experiments with a liquid crystal microdisplay *Eur. J. Phys.* **27**(5) 1221–31
[41] Boruah B R 2009 Dynamic manipulation of a laser beam using a liquid crystal spatial light modulator *Am. J. Phys.* **77**(4) 331
[42] Nicastro A J 1984 Experiment in critical phenomena at phase transitions in liquid crystals *Am. J. Phys.* **52**(10) 912–15
[43] Singh U and Moseley L L 1994 A simple experimental technique for investigating phase transitions using the DC Kerr effect *Eur. J. Phys.* **15** 154–6
[44] McConville S, Laurent D, Guarino A and Residori S 2005 Measurement of the giant nonlinear response of dye-doped liquid crystals *Am. J. Phys.* **73**(5) 425
[45] Orton J W 1987 Physics in an industrial research laboratory *Phys. Educ.* **22** 79–84
[46] Astin C, Talbot D and Goodhew P 2002 Weird materials *Phys. Educ.* **37**(6) 516
[47] Ogborn J 2004 Soft matter: food for thought *Phys. Educ.* **39**(1) 48–51
[48] Styer D F 2000 Insight into entropy *Am. J. Phys.* **68**(12) 1090–6

[49] Lambert F L 2002 Disorder—a cracked crutch for supporting entropy discussions *J. Chem. Educ.* **79**(6) 187–92

[50] Taylor P L and Tabachnik J 2013 Entropic forces—making the connection between mechanics and thermodynamics in an exactly soluble model *Eur. J. Phys.* **34**(3) 729–36

[51] Guyon E 1975 Second-order phase transitions: models and analogies *Am. J. Phys.* **43**(10) 877–81

[52] van Hecke G R 1976 Thermotropic liquid crystals *J. Chem. Educ.* **5**(3) 161–5

[53] Rosenbaum T F and Stein D L 1981 Topological physics illustrated in the laboratory *Am. J. Phys.* **49**(2) 128–33

[54] Sanati M and Saxena A 2003 Landau theory of domain walls for one-dimensional asymmetric potentials *Am. J. Phys.* **71**(10) 1005–12

www.ingramcontent.com/pod-product-compliance
Lightning Source LLC
LaVergne TN
LVHW081519300926
849161LV00006B/1219

* 9 7 8 1 6 4 3 2 7 8 3 7 7 *